Bioremediation of MTBE, Alcohols, and Ethers

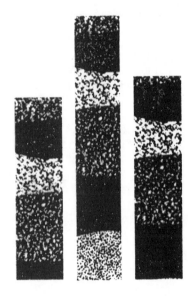

Editors

Victor S. Magar,
James T. Gibbs,
Kirk T. O'Reilly, Michael R. Hyman, and Andrea Leeson

The Sixth International In Situ and On-Site Bioremediation Symposium

San Diego, California, June 4–7, 2001

BATTELLE PRESS
Columbus • Richland

Library of Congress Cataloging-in-Publication Data

International In Situ and On-Site Bioremediation Symposium (6th : 2001 : San Diego, Calif.)
  Bioremediation of MTBE, alcohols, and ethers : the Sixth International In Situ and On-Site Bioremediation Symposium : San Diego, California, June 4-7, 2001 / editors, V. Magar ... [et al.].
     p. cm. – (The Sixth International In Situ and On-Site Bioremediation Symposium ; 1)
  Includes bibliographical references and index.
  ISBN 1-57477-111-6 (hc. : alk. paper)
   1. Butyl methyl ether--Biodegradation--Congresses. 2. Oxygenated gasoline--Biodegradation--Congresses. 3. Bioremediation--Congresses. I. Magar, V. (Victor), 1964- . II. Title. III. Series: International In Situ and On-Site Bioremediation Symposium (6th : 2001 : San Diego, Calif.). Sixth International In Situ and On-Site Bioremediation Symposium ; 1.
  TD192.5.I56 2001 vol. 1
  [TD196.P4]
  628.5 s--dc21
  [628.5'2]

2001044133

Printed in the United States of America

Copyright © 2001 Battelle Memorial Institute. All rights reserved. This document, or parts thereof, may not be reproduced in any form without the written permission of Battelle Memorial Institute.

Battelle Press
505 King Avenue
Columbus, Ohio 43201, USA
614-424-6393 or 1-800-451-3543
Fax: 1-614-424-3819
Internet: press@battelle.org
Website: www.battelle.org/bookstore

For information on future environmental conferences, write to:
  Battelle
  Environmental Restoration Department, Room 10-123B
  505 King Avenue
  Columbus, Ohio 43201-2693
  Phone: 614-424-7604
  Fax: 614-424-3667
  Website: www.battelle.org/conferences

CONTENTS

Foreword   vii

Biobarriers and Characterization of MTBE-Degrading Cultures

What Do Microbial Biosensors Tell Us About the Behavior of MTBE? *H. Maciel and G.I. Paton*   1

Demonstration of the Bioremedy Process for MTBE Remediation at Retail Gasoline Stations. *G.E. Spinnler, P.M. Maner, J.P. Salanitro, and P.C. Johnson*   11

Characterization and Degradation Kinetics of Aerobic MTBE Degrading Cultures. *A. Pruden, G.J. Wilson, M.T. Suidan, M.A. Sedran, and A.D. Venosa*   19

Natural Attenuation of MTBE

Natural Attenuation of MTBE as Part of an Integrated Remediation Design. *E.N. Moeri, M.C. Salvador, and R. Coelho*   27

Redox Conditions at Fuel Oxygenate Release Sites. *P. McLoughlin, R. Pirkle, T. Buschek, R. Kolhatkar, and N. Novick*   35

Natural Biodegradation of MTBE at a Site on Long Island, NY. *R. Kolhatkar, J. Wilson, and G. Hinshalwood*   43

Impact of Natural Attenuation and Phytoremediation on MTBE and Fuel. *K. Brown, L. Tyner, D. Grainger, T. Perina, M. Leavitt, and M. McElligott*   51

Natural Attenuation of MTBE in a Dual Porosity Aquifer. *G.P. Wealthall, S.F. Thornton, and D.N. Lerner*   59

Relative Depletion Rates of MTBE, Benzene, and Xylene from Smear Zone Non-Aqueous Phase Liquid. *T.R. Peargin*   67

Natural Attenuation of Dissolved Benzene and MTBE - Two Case Studies. *J. Robb and E. Moyer*   75

Hydrolysis of MTBE: Implications for Anaerobic and Abiotic Natural Attenuation. *K.T. O'Reilly, M.E. Moir, C. Taylor, and M.R. Hyman*   83

Ex Situ Bioreactors and Bench-Scale Studies

Fluidized Bed Bioreactor for MTBE and TBA in Water. *J.E. O'Connell and D. Weaver*   91

Enhanced Bioremediation of MTBE in Groundwater. *M.T. Balba, D. Coons,
R. Hoag, C. Lin, S. Scrocchi, and A. Weston* 99

In Situ MTBE Biodegradation

Aerobic Bioremediation of MTBE and BTEX at a USCG Facility. *P. Hicks,
M.R. Pahr, J.P. Messier, and R. Gillespie* 107

Field Data Regarding Air-Based Remediation of MTBE. *B.G. Billings and
J.E. Griswold* 115

Full-Scale Remediation of an MTBE-Contaminated Site. *J.J. Kang,
P.B. Harrison, and M.F. Pisarik* 121

Cost of MTBE Remediation. *B.H. Wilson, H. Shen, D. Pope, and S. Schmelling* 129

MTBE Cometabolism

Bioremediation of MTBE Through Aerobic Biodegradation and Cometabolism.
K.E. Hartzell, V.S. Magar, J.T. Gibbs, E.A. Foote, and C.D. Burton 137

Cometabolism of MTBE by an Aromatic Hydrocarbon-Oxidizing Bacterium.
M. Hyman, C. Smith, and K. O'Reilly 145

Selection of a Defined Mixed Culture for MTBE Mineralization. *A. Francois,
P. Piveteau, F. Fayolle, R. Marchal, P. Beguin, and F. Monot* 153

Biodegradation of MTBE and Other Gasoline Oxygenates by Butane-Utilizing
Microorganisms. *S.-W. Chang, S.-S. Baek, and S.-J. Lee* 161

Ethanol and Methanol Degradation

Impact of Ethanol on Benzene Migration. *H. Shen, S.C. Mravik, J.T. Wilson, and
G.W. Sewell* 167

In Situ Remediation of Ethanol, Ammonia, and Petroleum Hydrocarbons Via
Aerobic Biodegradation. *M.D. Nelson, B.W. Koons, B.J Peschong, C. Boben* 175

Methanol Behavior in the Subsurface at a Coastal Plain Release Site.
H.J. Reisinger, J.B. Raming, and A.J. Hayes 183

Effects of Ethanol Versus MTBE on BTEX Natural Attenuation.
M.L.B. da Silva, G.M. Ruiz, J.M. Fernandez, H.R. Beller, and P.J.J. Alvarez 195

Ethanol in Groundwater at a Northwest Terminal. *T.E. Buscheck, K. O'Reilly,
G. Koschal, and G. O'Regan* 203

Author Index 211

Keyword Index 239

FOREWORD

The papers in this volume correspond to presentations made at the Sixth International In Situ and On-Site Bioremediation Symposium (San Diego, California, June 4-7 2001). The program included approximately 600 presentations in 50 sessions on a variety of bioremediation and supporting technologies used for a wide range of contaminants.

This volume focuses on *Bioremediation of MTBE, Alcohols, and Ethers*. Environmental contamination by MTBE and other petroleum oxygenates, such as alcohols and ethers, has gained increasing and widespread attention over the last six to ten years. The first appearance of MTBE in the Bioremediation Symposium proceedings was in 1993, with the publication of four manuscripts. The 6th International In Situ and On Site Bioremediation Symposium saw four sessions with 40 platform and poster presentations dedicated to MTBE, and 26 manuscripts from those presentations are included in this volume.

This volume also covers the degradation of alcohols and ethers in pure form, or mixed with petroleum fuels from UST leaks. As MTBE use becomes increasingly regulated, alternative fuel oxygenates, such as alcohols and ethers are likely to be pursued. The increased use of alcohols and ethers amplifies their potential release into the environment through accidental spills or leaking underground storage tanks (UST).

The author of each presentation accepted for the symposium program was invited to prepare an eight-page paper. According to its topic, each paper received was tentatively assigned to one of ten volumes and subsequently was reviewed by the editors of that volume and by the Symposium chairs. We appreciate the significant commitment of time by the volume editors, each of whom reviewed as many as 40 papers. The result of the review was that 352 papers were accepted for publication and assembled into the following ten volumes:

Bioremediation of MTBE, Alcohols, and Ethers — 6(1). Eds: Victor S. Magar, James T. Gibbs, Kirk T. O'Reilly, Michael R. Hyman, and Andrea Leeson.

Natural Attenuation of Environmental Contaminants — 6(2). Eds: Andrea Leeson, Mark E. Kelley, Hanadi S. Rifai, and Victor S. Magar.

Bioremediation of Energetics, Phenolics, and Polycyclic Aromatic Hydrocarbons — 6(3). Eds: Victor S. Magar, Glenn Johnson, Say Kee Ong, and Andrea Leeson.

Innovative Methods in Support of Bioremediation — 6(4). Eds: Victor S. Magar, Timothy M. Vogel, C. Marjorie Aelion, and Andrea Leeson.

Phytoremediation, Wetlands, and Sediments — 6(5). Eds: Andrea Leeson, Eric A. Foote, M. Katherine Banks, and Victor S. Magar.

Ex Situ Biological Treatment Technologies — 6(6). Eds: Victor S. Magar, F. Michael von Fahnestock, and Andrea Leeson.

Anaerobic Degradation of Chlorinated Solvents — 6(7). Eds: Victor S. Magar, Donna E. Fennell, Jeffrey J. Morse, Bruce C. Alleman, and Andrea Leeson.

Bioaugmentation, Biobarriers, and Biogeochemistry — 6(8). Eds: Andrea Leeson, Bruce C. Alleman, Pedro J. Alvarez, and Victor S. Magar.

Bioremediation of Inorganic Compounds — 6(9). Eds: Andrea Leeson, Brent M. Peyton, Jeffrey L. Means, and Victor S. Magar.

In Situ Aeration and Aerobic Remediation — 6(10). Eds: Andrea Leeson, Paul C. Johnson, Robert E. Hinchee, Lewis Semprini, and Victor S. Magar.

In addition to the volume editors, we would like to thank the Battelle staff who assembled the ten volumes and prepared them for printing: Lori Helsel, Carol Young, Loretta Bahn, Regina Lynch, and Gina Melaragno. Joseph Sheldrick, manager of Battelle Press, provided valuable production-planning advice and coordinated with the printer; he and Gar Dingess designed the covers.

The Bioremediation Symposium is sponsored and organized by Battelle Memorial Institute, with the assistance of a number of environmental remediation organizations. In 2001, the following co-sponsors made financial contributions toward the Symposium:

Geomatrix Consultants, Inc.
The IT Group, Inc.
Parsons
Regenesis
U.S. Air Force Center for Environmental Excellence (AFCEE)
U.S. Naval Facilities Engineering Command (NAVFAC)

Additional participating organizations assisted with distribution of information about the Symposium:

Ajou University, College of Engineering
American Petroleum Institute
Asian Institute of Technology
National Center for Integrated Bioremediation Research & Development (University of Michigan)
U.S. Air Force Research Laboratory, Air Expeditionary Forces Technologies Division
U.S. Environmental Protection Agency
Western Region Hazardous Substance Research Center (Stanford University and Oregon State University)

Although the technical review provided guidance to the authors to help clarify their presentations, the materials in these volumes ultimately represent the authors' results and interpretations. The support provided to the Symposium by Battelle, the co-sponsors, and the participating organizations should not be construed as their endorsement of the content of these volumes.

Andrea Leeson & Victor Magar, Battelle
2001 Bioremediation Symposium Co-Chairs

WHAT DO MICROBIAL BIOSENSORS TELL US ABOUT THE BEHAVIOR OF MTBE?

Helena Maciel, Graeme I. Paton (Aberdeen University, Aberdeen, UK)

ABSTRACT: Although highly soluble and ubiquitous in the terrestrial environment there is little known about the impact of MTBE on microbial ecotoxicity tests. Microbial ecotoxicity tests are important as they are rapid indicators of environmental health and are also able to diagnose constraints to contaminant mineralisation. MTBE is rarely found in isolation and may have effects in combination with other hydrocarbons. To investigate the impact of MTBE on microbial ecotoxicity tests, the response of a range of *lux*-based biosensors was measured against aqueous samples of MTBE, benzene, and naphthalene, both individually and in combination. Metabolic sensors (indicative of general environmental stress) and catabolic sensors (induced by specific analytes) were used and the response modeled, enabling pollutant interactions to be identified. All concentrations were confirmed by PTI-GC-FID or HPLC. MTBE did not increase the toxicity of benzene or naphthalene to two of the sensors used but may have altered the pollutant bioavailability and indeed for some analytes reduced toxicity significantly.

INTRODUCTION

The storage of hydrocarbons and their widespread distribution has raised concern of their release into the environment. There is a growing need to assess the environmental fate and toxicity of hydrocarbons both in crude and refined oils. Additionally it is important that the behavior of fuel additives is effectively considered.

Methyl tertiary butyl ether (MTBE) is a fuel additive. It is produced in large amounts and used almost exclusively at post refinery stage. The widespread use of MTBE has resulted in surface and ground water contamination. MTBE is the one of the most frequently detected VOCs in ground water in urban areas though generally at low concentrations (Delzer *et al.*, 1996). The degradation half-life of MTBE is poorly documented (Vance, 1998).

Toxicity. There is a lack of information available regarding MTBE toxicity and often the information is contradictory. In the environment MTBE seldom appears alone hence it is important to study the interactions of MTBE with other gasoline compounds.

Pollutant mixtures can impose toxicity in several modes:
Additive: Toxicity of the mixture is equal to the sum of the individual components;
Synergistic: Toxicity of the mixture is greater than the sum of the individual components;

Antagonistic: Toxicity of the mixture is less than the sum of individual components.

The predicted concentration of the mixture (MTBE:benzene or MTBE:naphthalene) is calculated by the following equation, based on the luminescence response of the 2 individual compounds, equation 1.

(1)

$$\%lum_M/100 = (\%lum_A/100) * (\%lum_B/100)$$

Where M = the % of luminescence of the combination of two compounds
A = the % of luminescence of an individual compound
B = is the % of luminescence of the other compound being tested for interaction

The predicted value is then compared with the observed luminescence values of the mixtures, synergistic effects are when the observed values are significantly lower than the predicted values. If the reverse happens then the mixture is considered antagonistic. Paired t-tests were performed for all the combinations to evaluate interactions (Preston, 2000).

Hypothesis. Common gasoline components alter the toxicity of MTBE to specific receptors.

MATERIALS AND METHODS

Chemicals. MTBE and naphthalene were purchased from Sigma-Aldrich, UK, benzene was purchased from BDH (AnalaR). All standards were prepared in double deionised water and kept in Wheaton vials and stored without headspace at 4°C.

Two mL of the volatile organic compound (VOC) being prepared was added to 10 mL of double deionised water (ddH$_2$O) and shaken for one hour on an end over shaker. The Wheaton vials were inverted for half an hour and the soluble fraction of the VOC was extracted with a syringe. This was considered to be the stock solution. All the dilutions were made from the stock solution and the diluted standards were used immediately (or maintained without headspace and stored at 4°C).

The naphthalene saturated solution was prepared by mixing 20 mg of naphthalene crystals in 500 mL of ddH$_2$O.

Toxicity Testing. Effective concentration (EC$_{50}$) were determined for MTBE and 3 *lux*-marked bacterial biosensors and one *luc*-marked yeast (*Saccharomyces cerevisae*)

A range of concentrations of MTBE was tested against 3 freeze-dried biosensors (*E.coli* DH1, *E.coli* HB101 pUCD607 and *P.fluorescens* 10586r pUCD607) and a batch culture of yeast (*Saccharomyces cerevisae*).

E.coli HB101 pUCD607 and *E.coli* DH1 pUCD607 were resuscitated under aseptic conditions, by transferring 1 mL 0.1 M of potassium chloride (KCl), from a 10 mL vial, to the freeze-dried *E.coli* cells. After the cells were transferred to the universal bottle containing KCl and placed in an orbital shaking incubator at 200 rpm at 25 °C for 1 h. After that time the cells were used to perform the bioassay. A nine hundred μL aliquot of each sample (or standard) was pipetted into different cuvettes. One hundred μL of resuscitated cells were added to all the cuvettes with 15 seconds of interval. After 15 minutes the luminescence was measured using a portable luminometer. Luminescence was expressed as a percentage of the control (dd H_2O).

P.fluorescens, was resuscitated in Luria Bertoni (LB) and incubated at 25 °C for 1 h and transferred to sterile microcentrifuge tubes and centrifuged at 8720 g for 1 minute. The cells were ressuspended in 5 mL of 0.1 M KCl and the bioassay was performed as for *E.coli*. The yeast bioassay involved the use of batch cultures as described by Hollis *et. al.*, 2000.

Interactions. Mixtures of MTBE:Benzene and MTBE:Naphthalene were prepared between 75 and 1000 mg/L for MTBE; 25 and 800 mg/L for benzene and 7.5 to 25 mg/L for naphthalene.

The combination of different concentrations of MTBE and benzene to study interactions were determined using 6 concentrations at values around the EC_{50} and the bioassay performed as before.

Catabolic Biosensor. A batch culture of *P.fluorescens* HK44 was exposed for 1 h to several concentrations of MTBE, naphthalene, benzene (Burlage *et.al.*, 1994). For the combinations (MTBE:Benzene, MTBE:Naphthalene) the concentration of MTBE was always 1% v/v (51 mg/L).

Chemical Verification. The quantification of MTBE, benzene and naphthalene was carried out using a Purge and Trap Injection-Gas Chromatogram-Flame Ionization Detector or HPLC for naphthalene.

RESULTS AND DISCUSSION

Toxicity of MTBE. Figure 1 shows the response of the 4 metabolic *lux*-marked biosensors to MTBE. All values are normalized as a percentage of the control. A decrease in luminescence is observed with the increase the concentrations of MTBE, since increasing concentrations of MTBE become toxic to the biosensor. For the yeast there was no decrease in the light output for the range of concentrations tested. This biosensor does not seem to be affected by MTBE even at relatively high concentrations, such as 5000 mg/L (data not shown). *P.fluorescens* 10586r and *E.coli* DH1 seem to have a similar behavior when exposed to MTBE. *E.coli* HB101 seems to be less sensitive to MTBE and the EC_{50} is reached at higher concentrations.

The EC_{50} value for MTBE for the 3 biosensors seems to vary in a range, 200-600 mg/L however for benzene the EC_{50} is between 700-800 mg/L for the 3 biosensors. At low concentrations of benzene there is stimulation in luminescence expression (data not shown).

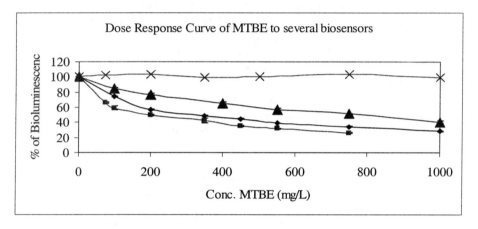

FIGURE 1. Percentage of bioluminescence for MTBE for the 4 *lux*-marked biosensors used: ♦ *P.fluorescens* 10586r and ■ *E.coli* DH1 and ▲ *E.coli* HB101 and × *Saccharomyces cerevisae*.

MTBE does impose acute toxicity on the cells tested. However, for the bacterial biosensors the EC_{50} is reached at a concentration much more elevated than that associated with environmental values. Further toxicity assays will be required with different organisms. Current work has progressed into the use of nematodes and earthworms this will be used to further investigate interactions.

Toxicity of MTBE:Benzene (metabolic biosensors). Using the model (equation 1), the observed luminescence values are higher than predicted for most of the combinations for *E.coli* DH1 and *P.fluorescens* 10586r. The opposite occurred for *E.coli* HB101, whose predicted toxicity was higher than observed when low concentrations benzene and MTBE were combined (figure 3).

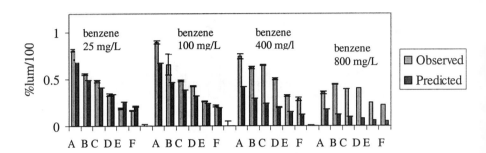

FIGURE 2. Evaluation of the interactions of benzene with MTBE. A =75, B= 200, C= 350, D= 500, E=750 and F=1000 mg/L of MTBE for *E.coli* DH1 both the predicted and observed values

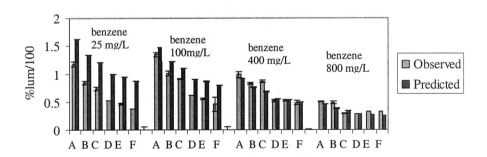

FIGURE 3. Evaluation of the interactions of benzene with MTBE. A =75, B= 200, C= 350, D= 500, E=750 and F=1000 mg/L of MTBE for *E.coli* HB101 both the predicted and observed values

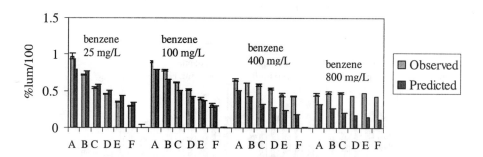

FIGURE 4. Evaluation of the interactions of benzene with MTBE. A =75, B= 200, C= 350, D= 500,E=750 and F=1000 mg/L of MTBE for *P.fluorescens* 10586r both the predicted and observed values

Based on the model, when MTBE solutions are combined with benzene (for *E.coli* DH1) for concentrations of benzene of 25 and 100 mg/L their effect is additive. Together or independently toxicity is not significantly different. For higher concentrations of benzene, there is a decrease of combined toxicity (antagonistic effect).

E.coli HB101 pUCD607, gave a different response: For low concentrations of benzene, there was an increase in toxicity, synergistic effect. But when the concentrations of benzene increased the combined toxicity started to decrease to values similar to the observed, additive effect.

P.fluorescens had a very similar response to *E.coli* DH1. For higher concentrations of benzene the combined toxicity value was significantly lower than the predicted.

Some of the observed differences were very small and appear insignificant.

Observed values were significantly higher than predicted values for *E.coli* DH1 and *P.fluorescens* 10586r, suggesting an antagonistic effect for high concentrations of benzene. For *E.coli* HB101 there wasn't significant difference between the observed and predicted value for high concentrations of benzene so the mixture toxicity was considered to be additive. Although for lower concentrations of benzene (25 and 100 mg/l) the predicted value was greater than the observed, hence exhibited a synergistic effect.

Two of the biosensors, *E.coli* DH1 and *P.fluorescens* 10586r seem to have a similar toxicological response to MTBE:Benzene while *E.coli* HB101 responded very differently to the others sensors. The combination of MTBE:Benzene was more toxic to this biosensor than when these pollutants are present alone (synergistic effect).

Toxicity of MTBE:Naphthalene (metabolic biosensors). For most of these combinations used there was no significant difference between the predicted and observed value.

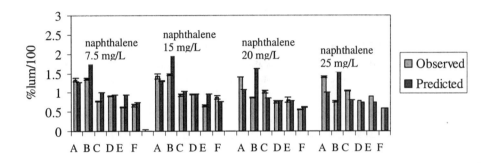

FIGURE 5. Evaluation of the interactions of naphthalene with MTBE. A =75, B= 200, C= 350, D= 500, E=750 and F=1000 mg/L of MTBE for *E.coli* HB101 both the predicted and observed values

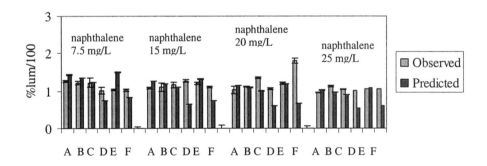

FIGURE 6. Evaluation of the interactions of naphthalene with MTBE. A =75, B= 200, C= 350, D= 500, E=750 and F=1000 mg/L of MTBE *P.fluorescens* 10586r both the predicted and observed values

For the mixture of MTBE:Naphthalene,toxicity was mostly additive. This may be due to the low aqueous solubility of napthahlene.

When MTBE is present with other pollutants such as benzene or naphthalene it seems that the toxic effects for most of the concentrations and biosensors are reduced hence a decrease in luminescence is observed. Current work is using soil microcosms to assess temporal and spatial interactions as a more environmentally relevant resolution.

Toxicity of MTBE (catabolic biosensor). The concentration of MTBE was 1% in all combinations. This concentration was chosen because for that concentration *P.fluorescens* HK44 emitted the highest amount of light in the presence of MTBE.

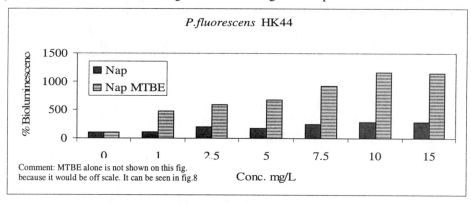

FIGURE 7. Evaluation of the effect of MTBE(1%):Naphthalene and naphthalene in HK44

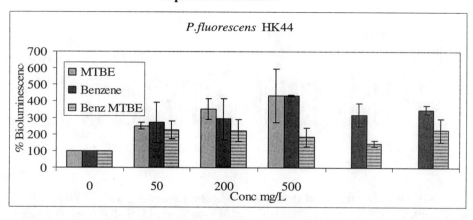

FIGURE 8. Evaluation of the effect of MTBE, MTBE(1%):Benzene and benzene in HK44

P.fluorescens HK44 is a specific catabolic biosensor to naphthalene. Unlike metabolic biosensors catabolic biosensors emit light when a specific inducer is present. For *P.fluorescens* HK44 it is the transformation of naphthalene to salicylate that causes luminescence to be emitted. The light emitted will be proportional to the inducer concentration. *P.fluorescens* HK44 is used to quantify naphthalene in contaminated sites. Bundy *et.al.*, 2000, showed that it also responds to other aromatic compounds like benzene. Figure 7 and 8 show that HK44 also responds to MTBE (50 to 500 mg/L) and also to the combinations of MTBE(1%):naphthalene and MTBE(1%):benzene. For MTBE(1%):naphthalene the induction of light is significantly higher than for naphthalene. The mechanism why this happens it is not clear. However dimethyl sulfoxide (DMSO), acetone and ethanol also stimulate the light output in *P.fluorescens* HK44 this might be indicative of some form of cellular disruption.

CONCLUSIONS

- MTBE is not acutely toxic to the microorganisms tested at environmentally relevant levels. This may mean that it posed little problem in the environment to microbial functionality. It seems to have an interactive effect with the aromatic compounds tested, in this study, although there is no conclusive evidence of an increase in toxicity.
- The relatively low level of sensitivity of the *lux*-based biosensors to MTBE does not reflect the environmental concern associated with this compound. The scope of work is being repeated using yeast, nematode and earthworms assays.
- Integrated approaches of microbial ecotoxicity and chemical analysis will enhance the understanding of environmental fate. The nature of MTBE and its association with other hydrocarbons means that it is particularly difficult to develop a comprehensive evaluation of the ecotoxicological characteristics.
- By characterizing spatial and temporal toxicity effects of MTBE we can enhance our ability to develop critical assessment tools.

ACKNOWLEDGMENTS

The authors would like to thank the Fundacao para a Ciencia e Tecnologia (Portuguese Ministry of Science) for funding this work and Dr. G. Sayler for supplying *P.fluorescens* HK44.

REFERENCES

Bundy, J.G., D.G. Durham, G.I. Paton, and C.D. Campbell. 2000. "Investigating the specificity of regulators of degradation of hydrocarbons and hydrocarbon-based compounds using structure-activity relationships". *BIODEGRADATION.* 11: (1) 37-47.

Burlage, R.S., A.V.Palumbo, A. Heitzer, and G. Sayler. 1994. "Bioluminescent reporter bacteria detect contaminants in soil samples" *Applied Biochemistry and Biotechnology.* 45-6: 731-740

Delzer, G.C., J.S. Zogorski, T.J. Lopes, and R.L. Bosshart. 1996. *Occurrence of gasoline oxygenate MTBE and BTEX compounds in urban stormwater in the United Sates 1991-95.* Prepared for United States geological Survey (Water-Resources Investigations report).

Drescher, K., and W. Bodeker. 1995. "Concept for the assessment of combined effect of substances: The relationship between concentration addition and independent action." *Biometrics* 51: 716:730.

Hollis, RP., K. Killham, and L.A. Glover. 2000. " Design and application of a biosensor for monitoring toxicity of compounds to eukaryotes". *Appl. Environ. Microbiol.* 66: (4) 1676-1679.

Preston, S., N. Coad, J. Towend, K. Killham, and G.I. Paton. 2000. "Biosensing acute toxicity of metal interactions: Are they additive, synergistic or antagonistic." *Env Toxicol Chem* 19: 775-780.

Stratton, GW. 1988. "Method for the determining toxicant interaction effects towards microorganisms." *Toxic Assess.* 3: 345-353.

Squillace, P.J., J.F. Pankow, N.E. Kortes, and J.S. Zogorski. 1997. "Review of the environmental behavior and fate of methyl tert-butyl ether." *Environmental Toxicology and Chemistry.* 16(9): 1836-1844.

Squillace, P.J., J.F. Pankow, N.E. Kortes, and J.S. Zogorski. 1998. " Environmental behavior and fate of methyl tertiary-butyl ether." *Water online newsletter.* 2(2)

Vance, D.B. *2 The 4 Technology Solutions.* (net).

Zogorski, J.S., A. Morduchowitz, A.L. Baehr, B.J. Bauman, D.L. Conrad, R.T.Drew, N.E. Korte, W.W. Lapham, and J.F. Pankow. "Fuel oxygenates and water quality: Current understanding of sources, occurrence in natural waters, environmental behavior, fate and significance."(net)

DEMONSTRATION OF THE BIOREMEDY PROCESS FOR MTBE REMEDIATION AT RETAIL GASOLINE STATIONS

G E. Spinnler, P. M. Maner and J. P. Salanitro
(Equilon Enterprises LLC, Houston, TX)
P.C. Johnson (Arizona State University, Tempe, AZ)

Abstract: MTBE remediation at retail gasoline stations has been demonstrated using an enhanced bioremediation system (BioRemedy). A specialized MTBE-degrading mixed culture was added to aquifer material in combination with oxygen creating bio-barriers to limit migration of MTBE plumes. MTBE is transformed to CO_2 as groundwater moves through the biobarrier. Results from these sites indicated up to four orders of magnitude reduction of MTBE concentrations within the biobarrier in a few months of operation. Biobarriers provide an effective means for controlling MTBE plume migration.

INTRODUCTION

MTBE remediation has proven difficult for conventional techniques. Biodegradation of MTBE by intrinsic organisms has not been well documented and anecdotal evidence of extremely long plumes suggests the process may not be widespread. Biostimulation of indigenous organisms, however, has been reported to occur in certain locations. With the addition of oxygen to aquifers, Salanitro et al. 1999, 2000 and Mackay et al., 2001 demonstrated apparent degradation of MTBE by naturally occurring aquifer microbes. MTBE degraders have not been observed to be as ubiquitous in the environment as BTEX degraders. Even when present, MTBE degraders may not have sufficient population density to degrade high concentrations of MTBE. Laboratory data indicate MTBE is an extremely poor growth substrate for the organisms, so rapid population growth may not occur in contaminated aquifers. Introducing MTBE-degrading microorganisms into aquifers is an alternative for biodegrading MTBE. Remediation of an MTBE plume using bioaugmentation has been successfully demonstrated at Port Hueneme, CA (Salanitro et al., 2000), and this technology has been used to control MTBE plume migration at several retail gasoline stations.

MATERIALS AND METHODS

Components of biobarrier system. Biobarrier system components are illustrated in Figure 1 and consist of MTBE-degrading microbes, an oxygen source and a monitoring well system. Soil amended with specialized MTBE-degrading microbes is located at or near the leading edge of the dissolved MTBE plume. Oxygen is supplied by pulsing oxygen gas through injection wells located within or near the microbe-rich zone. Figure 2 shows a map view of the system. With this design, groundwater contaminated with dissolved MTBE flows through the biobarrier and MTBE is degraded to CO_2 (Salanitro et al., 1998).

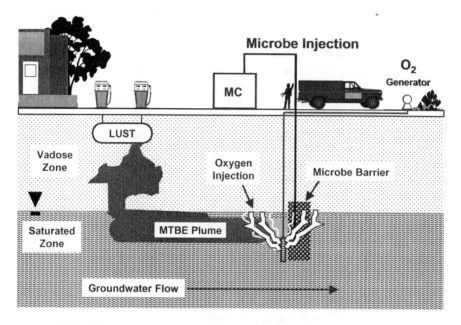

FIGURE 1. Components of the BioRemedy biobarrier system

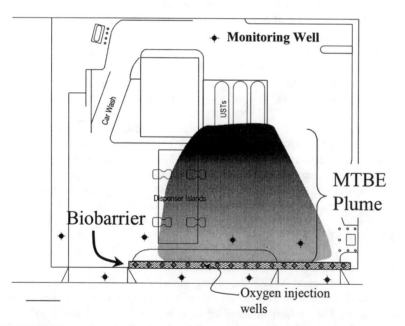

FIGURE 2. Map view of a biobarrier system installed at the edge of an MTBE plume.

Microbes. The MTBE-degrading culture is a proprietary aerobic mixed culture, designated MC (Salanitro et al., 1999). MC is capable of degrading MTBE, TBA as well as other common oxygenates. Results from laboratory microcosm experiments illustrate the effectiveness of using MC to degrade MTBE (Fig. 3). Soil from a retail site was used in a microcosm study to assess the presence of native MTBE degraders. MC was added to the soil at various loadings (mg MC/kg soil) to assess MTBE degradation rates. The native soil (without addition of MC) showed no MTBE degradation activity. All other soil samples degraded MTBE with their rates proportional to the MC loading.

MC was grown in commercial quantities at our facility in Houston, TX and shipped to sites. MC was injected into the aquifer using direct push techniques and a specially designed injection pump. Trenching can be used in cases where the soil characteristics make injection difficult or there is very limited permeability. At each site the vertical and horizontal extent of the injection zone was determined during site characterization.

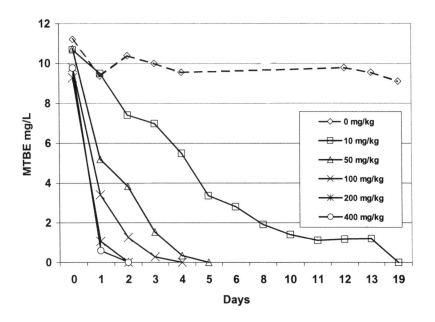

FIGURE 3. MTBE soil microcosm degradation.
MC concentration (mg/kg) in soil.

Oxygen. Since the transformation of MTBE to CO_2 is aerobic, oxygen is necessary in the biobarrier. Dissolved oxygen in the reaction zone is supplied using an oxygen pulse-sparge system. This system optimizes gas distribution and oxygen transfer while maintaining the permeability of the zone. Oxygen is generated on site and distributed using a solenoid-operated manifold system through injection wells. The vertical and horizontal well distributions were determined during field pilot tests. Gas injection wells were installed using conventional direct push techniques.

RESULTS AND DISCUSSION

Figure 4 is a map of a retail gas station located in the Northeast. MTBE levels up to 100 mg/L were reported in monitoring wells downgradient of the tanks and dispensers. Depth to groundwater was 13-15 ft. (4-5m) bgs. Soil consisted of indiscrete layers of sand, sand and gravel, and silty sand to a depth of approximately 30 ft. (10 m) bgs. BTEX concentrations were 1-5 mg/L if detected. A pilot test was conducted to assess injected gas distribution, microbe distribution, vertical extent of the MTBE plume and presence of native MTBE degrading organisms. No MTBE degrading activity was observed in microcosm studies. Oxygen-well placement, MC loading (mg of MC/kg soil) and microbe injection spacings were determined from data collected during the pilot test. Oxygen points were installed using direct push methods at two levels in the biobarrier due to gas stratification and the thickness of the MTBE plume. The oxygen system was operated for several weeks prior to MC injection to oxygenate the soil.

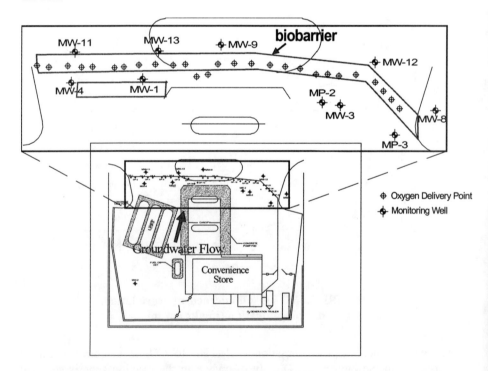

FIGURE 4. Retail gasoline station showing the approximate location of the biobarrier. Oxygen wells are located within the biobarrier.

MC was injected near the downgradient property boundary as indicated in Figure 4 using direct push tools and a specially designed high-pressure pump. Two rows of injections were installed in the area surrounding MW-1 due to higher MTBE concentrations. To ensure full coverage across the biobarrier, injections were made every 2 feet (.6 m) laterally and 2 feet (.6 m) vertically in the saturated zone.

Figure 5 is a plot of MTBE concentration versus time. The first data series was taken before the MC was installed. All other data series were collected after injection of MC. In all of the wells close to the biobarrier, i.e., within the injection zone, the MTBE concentration dropped, up to four orders of magnitude. MTBE concentration for wells MP-2 and SVE-1 were not observed to decrease over time (Fig. 5). Both MP-2 and SVE-1 are located upgradient from the biobarrier and indicate MTBE continues to enter the biobarrier.

Since no MTBE degradation activity was detected in the native soil, all MTBE degradation was determined to be due to the added MC.

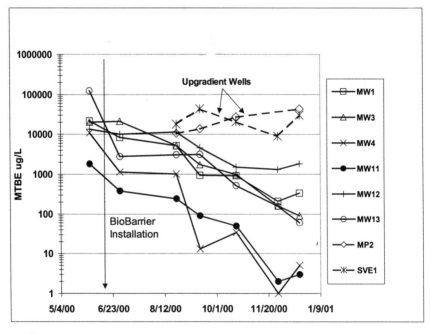

FIGURE 5. MTBE concentration versus time in monitoring wells within and upgradient of the biobarrier.

Biobarrier technology was also employed at a retail gas station in northern California (Figures 6a and b) to control offsite plume migration. MTBE concentrations had been >20 mg/L after tank excavation, and a sensitive receptor is located within a few hundred feet. Depth to groundwater at this site ranged from 3-5 ft. (1-2 m) bgs and an aquitard was located about 10 ft. (3 m) bgs. The

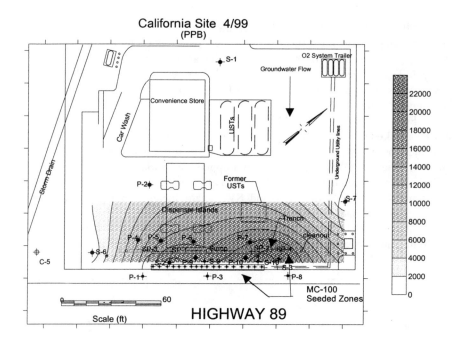

FIGURE 6a. MTBE Concentration before system installation.

FIGURE 6b. MTBE concentration after about one year of system operation.

soil consisted of a heterogeneous mixture of silty clay and gravel. Pilot testing indicated horizontal spacing of the oxygen wells. No MTBE degradation activity was observed in microcosm tests using native soil and groundwater.

A biobarrier was installed near the downgradient property boundary of this site to control offsite migration of the MTBE plume. Oxygen wells were installed using direct push methods and MC was emplaced by pressure injection. Oxygen is generated using an on-site oxygen generator system with manifold and solenoid control to pulse the oxygen wells. Oxygen is also supplied to the former groundwater extraction trench from the lateral collection pipe running along the base of the trench. The former groundwater extraction trench was also seeded to provide additional reduction capacity. Figures 6a and 6b show the location of the biobarrier, associated oxygen wells, the seeded trench and MTBE concentration isopleth maps. Figure 6a shows the initial concentration before the site was seeded with MC. Figure 6b shows the MTBE concentration after the system was operating about one year. Because no native degraders were present at the site, the dramatic decreases in MTBE concentration over a short period of time resulted from degradation due to the addition of MC in the biobarrier and the trench.

CONCLUSION

Utilizing MC as a biobarrier provides an effective MTBE remediation method for contaminated groundwater and a practical alternative to conventional methods. The wide occurrence of MTBE natural biodegradation has not been well documented. Because native MTBE degraders were not identified at the sites investigated, biostimulation was not a reasonable option. Adding a specialized mixed culture (MC) and oxygen to the aquifer, however, resulted in MTBE concentration reduction of up to four orders of magnitude due to biodegradation.

In situ treatment offers many advantages to conventional approaches for MTBE remediation such as pump and treat, while still achieving the goals of containment and mass reduction. MTBE is degraded to CO_2 underground, there is no need to dispose of water and bear the associated costs of pumping, treating and discharge and the possible cost of offgas treatment. Further, the cost to maintain the BioRemedy biobarrier system is expected to be less than a conventional pump and treat system since it has fewer components.

ACKNOWLEDGMENTS

We would like to thank Jennifer Bothwell of Equiva Services for her continued support of this work and John Hickey and staff of Geological Services Corporation, Hartford, CT for invaluable field assistance.

REFERENCES

Mackay, D., R. Wilson, K. Scow, M. Einarson, B. Fowler and I. Wood. 2001. *In Situ* remediation of MTBE at Vandenberg Air Force Base, California. *Soil Sediment and Groundwater, Special Edition* (In press).

Salanitro, J. P., C. C. Chou, H. L. Wisniewski, and R. E. Vipond. 1998. Perspectives on MTBE biodegradation and the potential for in situ aquifer remediation. Southwest Regional Conference of the National Groundwater Association, June 4-8, Anaheim, CA.

Salanitro, J., G. Spinnler, P. Maner, H. Wisneiwski, and P. C. Johnson. 1999. Potential for MTBE bioremediation-in situ inoculation of specialized cultures. Proceedings Petrol. Hydrocarbons and Org. Chem in Groundwater: Prevention, Detection and Remediation Conference, Nov. 17-20, Houston, TX.

Salanitro, J. P., P. C. Johnson, G. E. Spinnler, P. M. Maner, H. L. Wisniewski, and C. Bruce. 2000. Field-scale demonstration of enhanced MTBE bioremediation through aquifer bioaugmentation and oxygenation. *Environ. Sci. Technol.* 34:4152-4162.

CHARACTERIZATION AND DEGRADATION KINETICS OF AEROBIC MTBE DEGRADING CULTURES

Amy Pruden, Gregory J. Wilson, Makram T. Suidan, Marie A. Sedran
(University of Cincinnati, Cincinnati, Ohio)
Albert D. Venosa (U.S. EPA, Cincinnati, Ohio)

ABSTRACT: This study examines the effect of alternative substrate on the kinetics of methyl *tert*-butyl ether (MTBE) degradation. Four cultures were studied, each enriched under the following substrate conditions: MTBE with ethanol (EtOH), MTBE with diethyl ether (DEE), MTBE with diisopropyl ether (DIPE) and MTBE alone. Batch degradation experiments were carried out with each culture for both MTBE and *tert*-butyl alcohol (TBA), a key degradation intermediate of MTBE. A range of three initial concentrations (0.07 mM, 0.20 mM and 0.50 mM) were studied, with and without the addition of equivalent chemical oxygen demand (COD) of the alternative substrate. Each culture produced a unique degradation rate curve for MTBE, TBA, and overall degradation. The highest rates of degradation were observed in the MTBE only culture. The lowest rates of degradation were observed in the DIPE culture. The presence of an alternative substrate had an effect of slowing the rates of MTBE and TBA degradation, except in the case of DIPE, where it was observed to enhance the MTBE to TBA conversion step, but still slowed the overall degradation rate. Denaturing gradient gel electrophoresis (DGGE) was performed on all cultures and served in identifying the bacterial composition of the cultures, providing some explanation of the kinetic similarities and differences observed among the cultures.

INTRODUCTION
    MTBE contamination of groundwater is widespread throughout much of the U.S. as a result of its popular use in reformulated fuels. Most MTBE contamination is present in groundwater, originating from leaking underground storage tanks. MTBE is the most water soluble component of reformulated fuels, and therefore it forms plumes that spread very rapidly in the water table. It has now become a serious threat to drinking water in many communities, and it is currently classified by the U.S. EPA as a potential carcinogen. Biological treatment of MTBE contamination is a hopeful option, but has yet to be optimized in field applications. A better understanding of the factors that affect biological degradation of MTBE will aid in better application of bioremediation technologies.
    One pivotal issue in MTBE biodegradation is the role of alternative substrates. It is usually the case that MTBE is not the sole carbon substrate in contaminated groundwater. There are numerous reports that other water soluble gasoline components, such as benzene, toluene, ethyl benzene, and xylenes

(BTEX) interfere with the degradation of MTBE (Hubbard et al., 1994; Deeb et al., 2001). Alternatively, additional hydrocarbon substrates are sometimes intentionally added to the groundwater in order to stimulate cometabolic degradation of MTBE (Garnier et al., 1999).

In this study the effect of three alternative substrates on the rate of both MTBE and TBA degradation was examined. EtOH and DIPE, like MTBE, are used as fuel oxygenates in reformulated fuels, while DEE is an ether compound widely used as a solvent (See Table 1 for chemical structures of these compounds). Identification of the bacterial composition of the cultures provided insight into their kinetic behavior.

TABLE 1. Chemical Structures

Compound Name	Chemical Structure
Methyl *tert*-butyl ether (MTBE)	CH_3 $H_3C-\underset{\underset{CH_3}{\vert}}{\overset{\vert}{C}}-O-CH_3$
Diethyl ether (DEE)	$H_3C-CH_2-O-CH_2-CH_3$
Diisopropyl ether (DIPE)	$H_3C-\underset{CH_3}{\overset{\vert}{CH}}-O-\underset{CH_3}{\overset{\vert}{CH}}-CH_3$
Ethanol (EtOH)	H_3C-CH_2-OH
Tert-butyl alcohol (TBA)	CH_3 $H_3C-\underset{\underset{CH_3}{\vert}}{\overset{\vert}{C}}-O-H$

MATERIALS AND METHODS

Four aerobic cultures were enriched in identical chemostats, which contained a polyethylene porous pot for enhanced biomass retention, under the following substrate conditions: MTBE and DEE, MTBE and EtOH, MTBE and DIPE, and MTBE alone. The cultures were thus pre-acclimated to the substrate conditions of the batch tests to follow. Each culture was enriched with a continuous influent feed containing an equivalent total COD of 417 mg/L, which was evenly divided among both substrates for the dual-substrate cultures. The cultures were seeded with equivalent material: 2 liters of mixed liquor from the Metropolitan Sewer District (MSD), Cincinnati, Ohio, 600 mL of mixed liquor from Shell Co., Houston, Texas, and 140 mL of wash water of soil contaminated with MTBE provided by U.S. EPA-NRML.

Batch degradation experiments were carried out with each culture in 160 ml serum bottles containing 90 ml of chemostat effluent spiked with the appropriate amount of MTBE, TBA, or alternative substrate, and 10 ml of biomass from the corresponding chemostat. Sufficient bottles were set up for 7 triplicate sampling events. In addition, 4 mercuric chloride killed control bottles were sacrificed at beginning and at the end of the experiments, and provided for the time zero concentration point. Experiments were carried out in two batches for each culture (TBA batch followed by MTBE batch) and placed on the same tumbler with a rotational speed of 12 rpm. Degradation studies were done at three concentration levels: high (0.50 mM MTBE), medium (0.20 mM MTBE), and low (0.07 mM MTBE). This corresponds to about 45 mg/L, 17 mg/L, and 6 mg/L MTBE respectively. TBA studies were carried out at approximately equimolar concentrations as the MTBE batch, since MTBE conversion to TBA is also equimolar (Equation 1).

$$3/2\ O_2 + (CH_3)_3COCH_3\ (MTBE) \longrightarrow (CH_3)_3COH\ (TBA) + CO_2 + H_2O \quad (1)$$

$$6O_2 + (CH_3)_3COH\ (TBA) \longrightarrow 4CO_2 + 5\ H_2O \quad (2)$$

Equation 2 shows the subsequent mineralization of TBA. Alternative substrate, if present, was provided at an equivalent COD as the primary substrate (MTBE or TBA).

All concentrations were monitored through time with a Hewlett Packard 5890 Series II gas chromatograph (GC) (Hewlett Packard, Palo Alto, CA) equipped with a flame ionization detector (FID) and 60/80 Carbopack B5% Carbowax 20 M glass column (Supelco, Bellefonte, PA). Three rates were calculated for each culture in the following way (linear portion of curve only):

1) MTBE rate = [mmol $MTBE_{in}$ - mmol $MTBE_{out}$]/ min / mg VSS,
2) TBA rate = [mmol TBA_{in} - mmol TBA_{out}]/ min / mg VSS,
3) Overall rate = [mmol$(MTBE + TBA)_{in}$ – mmol$(MTBE + TBA)_{out}$]/ min / mg VSS.

VSS measurements were done with each batch in quadruplicate in order to determine the overall biomass concentrations. DGGE, a molecular tool for identifying bacterial communities was also carried out with each batch in order to determine the compositions of the cultures, as well as to confirm that the cultures remained consistent between batches. The primers used for DGGE were designed for universal detection of bacteria. See Muyzer et al. (1993) for details on DGGE methods.

RESULTS AND DISCUSSION

For all cultures, MTBE disappearance with time was faster than TBA disappearance with time. This is seen by comparing Figure 1 with Figure 2. Figure 1 is a plot of the rate of MTBE disappearance with time at each of the three initial concentrations, not considering TBA. Figure 2 shows TBA degradation rates for the TBA batch studies. The overall rate of MTBE degradation (Figure 3), defined here as complete mineralization of MTBE and intermediate TBA, exhibited comparable rates to those of TBA degradation alone. In addition, TBA was observed to build up during MTBE degradation before finally being degraded to completion. Figure 4 shows a curve illustrating TBA build-up that was typical of all four cultures. Thus there is strong evidence in support of the hypothesis that TBA degradation is the rate-limiting step in MTBE biodegradation.

The presence of an alternative substrate was observed to slow the degradation rate of both MTBE and TBA, as well as the overall degradation rate, with the exception of the DIPE culture (Figures 1-3). This suggests that either the same organisms are degrading both the MTBE/TBA and the alternative substrate, or that there is competition for some resource involved in degradation. DGGE analysis of the DEE culture revealed that only one organism (belonging to the analysis of the DEE culture revealed that only one organism (belonging to the *Flavobacteria Cytophaga* see Figure 6) dominated the culture at the time of the experiment, giving support to the former scenario. All other cultures consisted of more than one dominant organism detectable by DGGE and contained at least one member of the *Flavobacteria Cytophaga*. These organisms were observed to be associated with MTBE degradation in a previous study. In addition, the DIPE culture contained 2 members of the genus *Sphingomonas* (Fig. 6), which are known hydrocarbon degraders, and the EtOH culture contained 3 other members of the α-Proteobacteria.

The DIPE culture was the only one of the three that exhibited some evidence of cometabolism. When DIPE was present, the MTBE to TBA conversion rate was faster, but only for the low and intermediate concentrations (Figure 1, Figure 5). Nevertheless, the overall rate of MTBE degradation was still hindered by the presence of DIPE (Figure 3).

Comparison between cultures shows that the MTBE culture was capable of the fastest degradation rate of both MTBE and TBA, as well as the fastest overall rate (Figures 1 - 3). A likely explanation for this is that this culture is best acclimated to MTBE mineralization and has been able to channel its degradative resources solely towards the destruction of MTBE and its intermediate. This is further evidence that the presence of alternative substrate hinders the rate of MTBE degradation. In this case an alternative substrate elicits a decreased rate not only because of direct competition for the activity of the microorganisms, but because it affects the actual development of the degrading community.

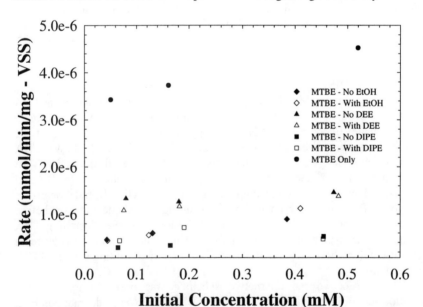

FIGURE 1. Plot of MTBE degradation rates corresponding to initial batch concentration of MTBE.

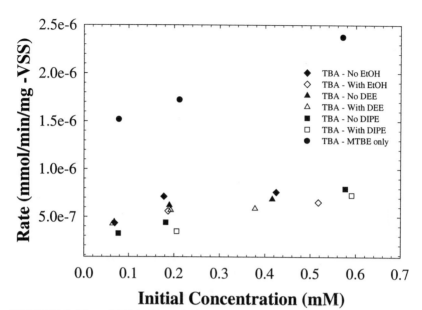

FIGURE 2. Plot of TBA degradation rates corresponding to initial batch concentration of TBA.

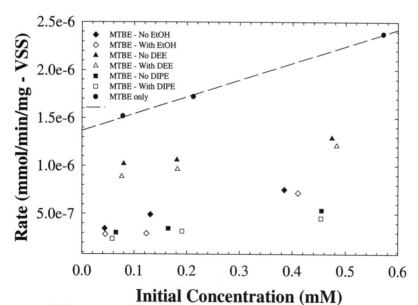

FIGURE 3. Plot of overall degradation rate of complete MTBE mineralization corresponding to initial batch concentration of MTBE

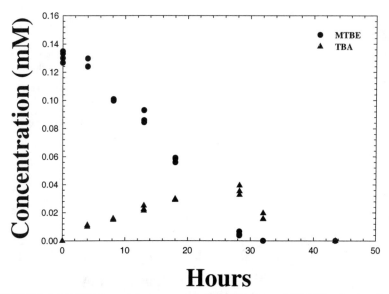

FIGURE 4. MTBE degradation curve for MTBE and EtOH culture showing characteristic accumulation of TBA observed in each of the four cultures

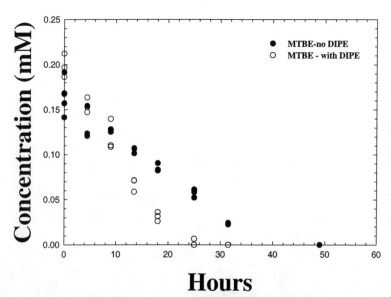

FIGURE 5. MTBE degradation curve for MTBE and DIPE culture 0.20 mM study. DIPE had a similar effect in the 0.07 mM study, but had no enhancing effect in the 0.50 mM study.

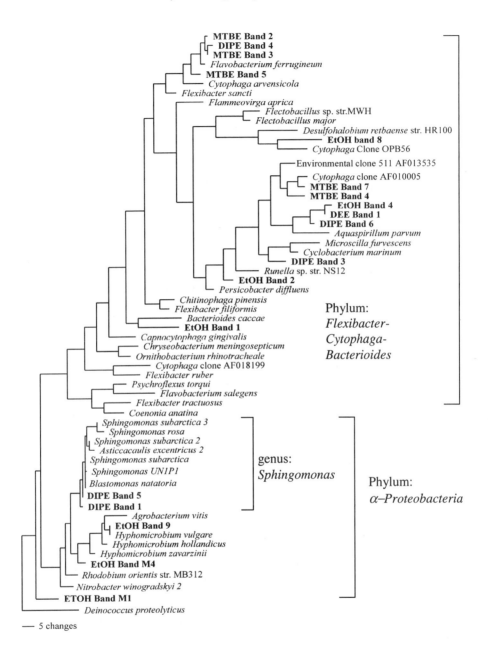

Figure 6. Phylogenetic analysis of bacteria detected by DGGE (shown in bold) generated by maximum likelihood analysis using PAUP*. The name of the band corresponds to the culture of origin. Names of organisms in italics represent organisms from the Ribosomal Database Project (RDP), shown here for evolutionary comparison.

The slowest degradation rates were observed in the DIPE culture. The reasons for this are subject to speculation, but the DGGE results confirm the presence of at least one genus of bacteria, *Sphingomonas*, which are not present in any of the other cultures (Fig. 6). This organism could be either interfering with MTBE degradation, or itself be degrading MTBE at a different rate, and thus possibly under a different mechanism than the degraders in the other cultures.

In this study it is seen that four aerobic cultures produce four unique degradation curves. DGGE results showed that the community composition of the cultures is also distinct. This diversity may be an indicator of multiple mechanisms, or enzyme pathways for MTBE biodegradation, which will be promising for flexible application in the field. An important similarity among the cultures was the apparent rate-limiting effect of TBA. This may prove to be a bottleneck in MTBE degradative pathways, and in a positive sense provide a focus for optimization in future research. Furthermore, it is seen that the presence of alternative substrate had a negative impact on the rate of overall MTBE degradation in all cases. The presence of alternative substrates can therefore be expected to slow MTBE degradation. By far the most efficient culture was the MTBE culture, which never had any exposure to an alternative substrate. These results may all contribute to improved design of MTBE bioremediation efforts, both *in situ* and *ex situ*.

ACKNOWLEDGMENTS

This research was made possible under Contract No. 68-C7-0057 from the U.S. Environmental Protection Agency. The findings and conclusions expressed in this paper are solely those of the authors and do not necessarily reflect the views of the Agency.

REFERENCES

Deeb, R. A., H-Y Hu, J. R. Hanson, K. M. Scow, and L. Alvarez-Cohen. 2001. "Substrate interactions in BTEX and MTBE mixtures by an MTBE-degrading isolate." *Environ. Sci. Technol.* 35(2):312-317.

Garnier, P. M., R. Auria, C. Augur, and S. Revah. 1999. "Cometabolic biodegradation of methyl t-butly ether by Pseudomonas aeruginosa grown on pentane." *Appl. Microbiol. Biotechnol.* 51:498-503.

Hubbard, C. E., J. F. Barker, S. F. O'Hannesin, M. Vandegriendt, and R. Gillham. 1994. *Transport and fate of dissolved methanol, methyl-tertiary-butyl-ether, and monoaromatic hydrocarbons in a shallow sand aquifer.* Publication 4601, p. 226. American Petroleum Institute, Health & Environmental Sciences Department, Washington, D.C.

Muyzer, G., E. C. de Waal, and A. G. Uitterlinden. 1993. "Profiling of microbial populations by denaturing gradient gel electrophoresis analysis of polymerase chain reaction amplified genes coding for 16S rRNA." *Appl Environ. Microbiol.* 59:695-700.

NATURAL ATTENUATION OF MTBE AS PART OF AN INTEGRATED REMEDIATION DESIGN

Ernesto Niklaus Moeri, Maria Cristina Salvador and Rodrigo Coelho
CSD-GEOKLOCK, São Paulo, Brazil
(E-Mail csdgeo@geoklock.com.br)

ABSTRACT

At a former industrial landfill site in southeastern Brazil, phenol and MTBE infiltrated into the soil until 1989. Significant concentrations of both pollutants have reached the local groundwater, and the contamination plume, currently over 400 meters long, is moving towards surface waters, thus putting at risk external receptors. The aquifer is shallow, has low hydraulic conductivity and a flow velocity of less than 20 meters per year. Water temperature is $25 - 30^{\circ}C$.

The remediation concept for the contaminated groundwater has been designed to ensure the protection of the surface water receptors, and consists of two hydraulic barriers that intercept the front end of the plume. Monitoring data available from a period of more than 10 years for phenol in groundwater show clear evidence for the existence of significant natural attenuation processes which are responsible for a dynamic equilibrium of this contaminant in the groundwater.

As for the MTBE concentrations, monitoring data have only been collected over the last two years and indicate a slower plume movement than predicted from the mathematical modeling with no decay factor. Detailed studies of groundwater chemistry along the contaminant flow path supplied further evidence of the existence of considerable degradation of MTBE, presence of aerobic bacteria, depletion of dissolved oxygen in the source zone of the plume and slightly increasing concentrations towards the fringe zone, increasing concentrations of specific MTBE degradation products such as TBA over the flow path, and reduction of total contained MTBE mass in the groundwater over time. A half-life of approximately 2.5 years for the decay of MTBE was obtained. This value is consistent with the values obtained in laboratory and pilot tests as reported in the literature.

The reactions responsible for the MTBE degradation required presence of oxygen. In order to maintain and enhance the rates of intrinsic bio-degradation observed at this site, complementary remediation technologies, such as advanced oxidation, were considered as promising tools to speed up the operational time for achieving the established remediation and clean-up goals.

SITE BACKGROUND

An industrial landfill site where phenol and MTBE have been infiltrated in open ponds was operated for over 20 years until 1989. The local groundwater has been monitored for more than 10 years for phenol and for a little more than 2 years for MTBE, using 47 permanent monitoring wells and several temporary sampling points (*Geoprobe®*) (**Figure 1**). The groundwater shows a radial flow pattern and is of shallow depth (10 – 20 meters). It has low hydraulic conductivity (around 10^{-5} cm/s) in the tropically weathered zone which consists of residual soil and very low hydraulic conductivity (less than 10^{-6} cm/s) within the underlying basaltic rocks of cretaceous age and volcanic origin. The temperature of the groundwater varies between 25 and 30°C.

Figure 2 shows a hydrogeological cross-section throughout the site where the gradual decrease of groundwater depth towards the natural outcrop can be seen.
Several phases of investigation have been carried out to provide a full characterization of the site in terms of vertical and lateral extension of the contamination, including geophysical surveys (ground penetrating radar) to define the bedrock morphology, physical, chemical and bacteriological characterization of the groundwater, as well as 3D flow and transport modeling. Most of the investigation has focused on the shallow portion of the aquifer, where the movement of pollutants preferably occurs. Downward migration of contaminants is limited through the very homogeneous, compact lithology of the local bedrock, which has been investigated through deep drill holes and specific permeability investigation (high pressure tests) along the rock sequence. Geochemical investigation of the former waste ditches and the underlying soil indicates that more than 90% of total MTBE mass has been transferred to the groundwater over the last 10 years, since the infiltration and deposition have been stopped.

CONTAMINATION CHARACTERISTICS

Five sampling and assaying campaigns provided considerable hydrochemical data to allow a comprehensive interpretation of the contaminant plume. Significant concentrations of both pollutants have reached the groundwater. Phenol concentrations vary from 4,237,000 μg/L to 0.95 μg/L and MTBE from 159,000 μg/L at the center of the plume to 1.10 μg/L at the leading edge. MTBE is visibly more persistent along the contaminant flow path, due to its higher relative solubility, and moves faster towards the surface water receptors. The plume delineation and the observed absolute concentrations show the influence caused by the seasonal variations of the groundwater heads, thus indicating a reduction of dissolved contaminant mass during the dry season (April – October), when groundwater levels are much lower than during the rainy season. **Figure 1** shows the general distribution of the major pollutant, MTBE, in the aquifer; it can be observed that "wetland 2", as one of the receptors, has already been reached. Mathematical modeling of MTBE transport in groundwater, without considering any active degradation mechanism, indicates that the plume would reach the second one ("wetland 1") within the next two years.

REMEDIATION CONCEPT

Based upon the above-mentioned data from the mathematical modeling, and considering the evidences of intrinsic attenuation processes, the remedial actions were designed in order to protect the external receptors (wetlands) within 30 µg/L for MTBE. The respective remediation goals have been discussed and agreed upon with the local environmental agency. In order to protect the receptors, two hydraulic barriers were engineered and installed at the fringe zone of the plume, to conform with the local hydraulic properties of the aquifer. For one barrier, a sequence of conventional vertical pumping wells was used, whereas for the "wetland 1" receptor, where hydraulic conductivity was extremely low, HDD (Horizontal Directional Drilling) technology was used and a total of 600 meters of horizontal filters installed (**Figure 1**). Both barriers had been completed by October 2000 and are showing good efficiency and performance according to the mathematical model. A complete drawdown of the groundwater was obtained 10 weeks from start-up. Total pumping rate is 5.5 m^3 per hour. The groundwater is pumped through a 2,500-meter pipeline to a treatment plant, located at the nearby industrial site.

The operational, hydraulic and environmental performance of the system is continuously monitored.

DISCUSSION OF MONITORING DATA

For all four complete monitoring campaigns the total contained MTBE mass in the aquifer was calculated and compared against total groundwater volume, considering seasonal variations of the saturated zone thickness. The porosity used was 45%. **Table1** shows the results obtained.

TABLE 1 - MASS COMPARISION ALONG TIME

	May/99	**October/99**	**June/00**	**January/01**
Contaminated area (1,000 m^2)	169	178	187	175
Volume of the impacted aquifer (1,000 m^3)	598	472	468	472
Mass of MTBE (Kg)	3,175	3,108	2,690	1,989

The graph on **Figure 3** shows the dissolved MTBE mass over time. Total mass reduces over time, however, with variations between dry and wet season. Mass reduction over two years reaches 37 %, indicating the existence of significant intrinsic attenuation processes. The observed mass reduction 12 years after the contaminant infiltration, is consistent with a possible acceleration of the degradation process as observed in the Borden Pilot Test (Schirmer and Weiss, 1999). It is important to note that mass removal at the hydraulic barriers does not exceed to the present date more than 13 kilograms, or less than 1% of total contained mass.

The above data indicate that considerable mass reduction occurs within the aquifer. In order to understand the possible chemical, physical and biological processes which are responsible for the attenuation, additional assays of groundwater observation points along the contaminant flow path have been carried out. A total of four monitoring wells, located in distinct areas within the contamination plume, have been sampled and assayed over time. The respective information obtained is presented in **Figure 2**. Along with the reduction of MTBE concentration, an increase in the aerobic bacteria population (10^0 to 10^4 ufc/ml), a relative increase in TBA, and a significant increase of dissolved oxygen from the source to the fringe zone were measured.

The appearance of TBA, considered a common intermediate degradation product of MTBE (Church et al., 1997), is a positive indication for MTBE biodegradation (Yeh and Novak, 1994). The observed concentrations along the flow path also indicate that TBA is sufficiently resistant to further degradation and that it may accumulate as an intermediate before further degradation (Salanitro et al., 1994; Mormile et al., 1994). The dissolved oxygen values and the observed bacteria population indicate aerobic biodegradation as the most probable and dominant process of MTBE removal.

The figure also shows that MTBE decay is more accelerated in the regions of the aquifer where phenol occurs in concentrations of the same order, decreasing sharply as the phenol is consumed.

Some further degradation products of TBA such as 2-propanol, acetone and CO_2 have also been detected. However, most of these products can also be derived from other contaminants.

CALCULATED RATE OF DECAY

The degradation rates of this site can be expressed in terms of a first order kinetic model. The monitoring data obtained along the flow path of contaminants were treated with the Bioscreen® software, that allows a quantification of the identified natural degradation processes. Through this model it was possible to adjust the values of natural degradation that approach the concentration data obtained in the field at each sampling campaign. An apparent half-life ($t_{1/2}$) of 2.5 years for the decay of MTBE at this site was obtained. The monitoring data and the obtained MTBE decay rate were used on an MT3D mathematical transport model to simulate the degradation process over time. The plume modeled by MT3D is very similar to what is observed in the field.

The rate of transformation of contaminants can be expressed by the equation: $C_t = C_0 e^{-kt}$, where C is the concentration of the pollutant and k is the apparent first-order rate constant (year^{-1}).

The MTBE mass over time was precisely calculated so it could be used as an alternative for the pollutant concentrations for the calculation of the decay rates. As a result for MTBE at this site the degradation rate obtained is 0.28 year^{-1} and the half-life is 2.5 years, as can be observed in **Figure 3**.

The MTBE half-life obtained in this study is quite solid when compared to the results presented in previous work.

Work based on laboratorial experiments (Church et al., 2000) obtained half-lives of conversion of MTBE for TBA of 2.7 years, 1.7 years and 1.9 years. Few studies are based on data observed in the field (Borden et al., 1997, Schirmer and Barker, 1998) and in that work the results obtained indicated a half-life of around 2 years. It stands out however, that in those studies the MTBE plume was always related to gasoline leaks, and a great part of the indicated data of natural reduction is conditioned to the biodegradation of other gasoline constituents.

CONCLUSIONS

The evidence of significant MTBE degradation in groundwater is of strong relevance for the simplification of the remedial design of a major contamination clean-up case. The calculated rates of decay can be considered as high as the ones reported from laboratory studies (Church, 1997; Borden et al., 1997; Schirmer and Barker, 1998) and similar to the reported degradation rates observed in pilot field tests. To our present knowledge the responsible mechanisms of degradation are related to biological activities of aerobic bacteria driven by the availability of oxygen and nutrients.

Based upon these data, there is considerable potential for reduction of the overall remediation time and costs. Further studies are being carried out with the objective of defining additional *in-situ* interventions, such as advanced oxidation, in order to enhance the naturally occurring processes.

ACKNOWLEDGEMENTS

The authors acknowledge the continuos support of CSD-GEOKLOCK staff, especially Daniela Goulios and Marcos Sillos and the critical review and analytical support of Michel Schurter and Rolf Gloor of Institut BACHEMA, Switzerland.

REFERENCES

Borden, R. C., R. A. Daniel, L. E. Le Brun, C. W. Davis. 1997. "Intrinsic Biodegradation Rates of MTBE and BTXE in a Gasoline-Contaminated Aquifer". *Water Resource. Res.* 33: 1105-1115.

Church, C. D., L. M. Isabellle, J.F. Pankow, D. L. Rose and P. G. Tratneyek. 1997. "Method for Determination of Methyl tert-Butyl Ether and its Degradation Products in Water". *Environmental Sci. Techn.* 31(12): 3723-3726.

Church, C. D., P. G. Tratnyek, J. F. Pankow, J. E. Landmeyer, A. L. Baehr, M. A. Thomas and M. Schirmer. 1999. "Effects of Environmental Conditions on MTBE Degradation in Model Column Aquifer". *U.S.G.S. Water Resources Investigations Report.* 3: 93-101.

Church, C. D., J. F. Pankow and P. G. Tratnyek. 2000. "Effects of Environmental Conditions on MTBE Degradation in Model Column Aquifers: II. Kinetics". *Environ. Chem. Div. Extended Abstracts*, 219th ACS Nat. Mtg. 40(1).

Gray, M. R., D. K. Banerjee, M. J. Dudas and M. A. Pickard. 2000. "Protocols to Enhance Bidegradation of Hydrocarbon Contaminants in Soil". *Battelle Memorial Institute Bioremediation Journal.* 4(4): 249-257.

Mormile, M. R., S. Liu and J. M. Suflita. 1994. "Anaerobic Biodegradation of Gasoline Oxygenate: Extrapolation of Information to Multiple Sites and Redox Conditions". *Environmental Sci. Techn.* 28(9): 1727-1732.

Salanitro, J. P., L. A. Diaz, M. P. Williams and H. L. Wisniewski. 1994. :"Isolation of a Bacterial Culture that Degrades Methyl t-Butyl Ether". *Applied and Environmental Microbiology.* 60(7): 2593-2596.

Schirmer, M. and J. F. Barker. 1998. "A Study of Long-Term MTBE Attenuation in the Borden Aquifer, Ontario, Canada". *Ground Water Monit. Rem.* 18: 113-122

Schirmer, M and H. Weiss. 1999. "Einfluss Refraktrarer Substanzen wie Methyl-tertiarether (MTBE) auf "Natural Attenuation" – Ansatz in Grundwasserleitern". *Altlasten Spektrum.* 6/99: 340-344.

Yeh, C. K. and J. T. Novak. 1994. "Anaerobic Biodegradation of Gasoline Oxygenates in Soils". *Water Environment Research.* 66(5): 744-752.

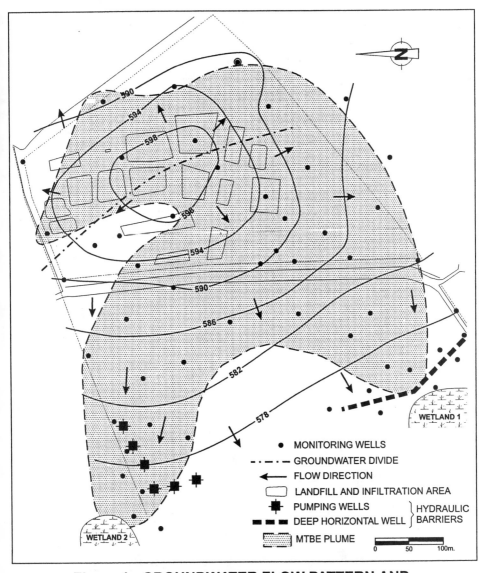

Figure 1 - GROUNDWATER FLOW PATTERN AND LAY OUT OF REMEDIAL ACTIONS

34   *Bioremediation of MTBE, Alcohols, and Ethers*

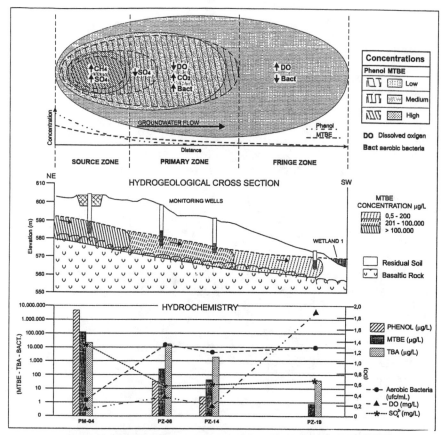

Figure 2 - INTRINSIC BIODEGRADATION PROCESS OF MTBE

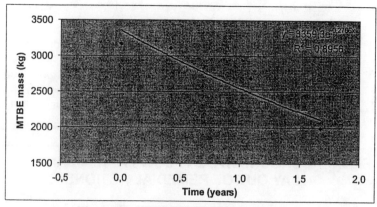

Figure 3 - MTBE MASS EVOLUTION

REDOX CONDITIONS AT FUEL OXYGENATE RELEASE SITES

Patrick McLoughlin, Robert Pirkle (Microseeps, Inc.)
Tim Buscheck (Chevron Research and Technology Company, Richmond, CA)
Ravi Kolhatkar (BP, Naperville, IL)
Norman Novick (ExxonMobil Corporation, Fairfax, VA)

Members of the Soil and Groundwater Technical Task Force of the American Petroleum Institute (API) have identified a number of service station sites on the East and West Coasts where, based on plume evaluations, natural attenuation of MTBE appeared to be occurring. In order to characterize redox conditions and identify the predominant metabolic processes, Microseeps field personnel sampled 18 of these gasoline stations made available by the API members BP, Chevron and ExxonMobil. The sites were in California (13), New Jersey (4) and Pennsylvania (1). Six wells were sampled at most of the sites. Analytical results are being compiled and Microseeps personnel are preparing an interpretive report.

A review of that project is presented, and includes 1) the screening criteria for choosing sites; 2) the analytical program that was chosen; 3) the sampling methodologies used; and 4) the preparation of data to evaluate the conditions graphically. Additionally, conclusions drawn from the overall review of the project are discussed. These conclusions will address the range of redox conditions found, the range of geochemistries surveyed and the role of controlling indicators. Several criteria to assess potential for natural attenuation of MTBE are discussed. Four different categories of sites are defined (*e.g.* methanogenic - sulfate-depleted) and assigned. A subset of the sites is recommended for microcosm studies to evaluate the field observation of MTBE natural attenuation.

INTRODUCTION

Methyl tert-butyl ether, (MTBE), is a gasoline oxygenate that has had both positive and negative impacts upon the environment (Davis et al., 1997;USEPA, 1999). The role of biodegradation as a loss mechanism for MTBE under various redox conditions is unclear. Past experience with hydrocarbon and chlorinated hydrocarbon plumes indicate that the rates and extent of natural biodegradation strongly depend on the prevalent redox conditions (Wiedemeier et al., 1998). Since the exact mechanisms of MTBE biodegradation are still being understood, two broad categories will be considered: those of oxic and anoxic conditions. Tert-butyl alcohol, (TBA), is a product of most degradation pathways that can readily undergo biodegradation through either aerobic (Steffan et al., 1997) or anaerobic (Hickman et al., 1989) processes. Aerobic MTBE biodegradation has been recently demonstrated as a viable remediation alternative at a number of research sites (Mackay et al., 1999; Barcelona and Jaglowski, 1999; Salanitro et al., 2000). Anaerobic biodegradation of MTBE is much less well understood. However, MTBE is typically released as part of reformulated gasoline and biodegradation of the less recalcitrant components of gasoline often leaves the MTBE in anoxic environments. Unless enhanced aerobic MTBE remediation is undertaken, it is only the anaerobic biodegradation processes that can substantially contribute to MTBE fate. A review of the anaerobic biodegradation literature was published in a paper by researchers from the US EPA's R. S. Kerr Environmental Research Center. That paper also presents the first field-scale evidence of the natural biodegradation of MTBE under methanogenic conditions. The field scale biodegradation rate constant was observed to be approximately 2.7 per year with laboratory confirmation of the biodegradation (Wilson et al., 2000). As further corroboration of that work, a more recent paper reported field evidence of natural biodegradation of MTBE (Kolhatkar et al., 2000).

The rate constant reported in the methanogenic study (Wilson et al., 2000) corresponds to a half-life of three months. A report of an MTBE contaminated groundwater remediation project that used in-situ air sparging and soil vapor extraction claimed a 50% MTBE removal in one year (Giattino et al., 2000). An active remedial effort that stimulated aerobic respiration shows 50% degradation achieved in 1-2 months after BTEX removal (Koenigsberg, et al. 1999).

The American Petroleum Institute (API) has undertaken a project designed to expand the database on, and understanding of, natural MTBE biodegradation. The project objectives are to: (1)Demonstrate MTBE biodegradation in aquifer materials under appropriate redox conditions; (2)Better evaluate the frequency with which MTBE biodegradation will be found in the saturated zone; (3)Evaluate whether rates of MTBE biodegradation, where found, are adequate to affect plume size; and (4)Correlate MTBE biodegradation with geochemical indicators and/or site characteristics (e.g., age of MTBE release). The ultimate objective of the project is to build the basis for an MTBE natural attenuation protocol similar to those available for other contaminants (Wiedemeier, et al., 1995; ASTM, 1998; Wiedemeier et al., 1998).

Three API member companies, BP, ExxonMobil Corporation and Chevron Research and Technology Company participated in the program. The effort consists of three phases: I) site selection II) geochemical site characterization and III) microbiological laboratory studies. Representatives of the participating petroleum producers carried out Phase I. API enlisted Microseeps to perform the sampling, analysis and data review of Phase II. Microseeps provided input concerning sampling feasibility, etc. to the selection process, but the principle criteria for site selection were as follows:

- No ongoing active remediation activities
- Adequate spatial MTBE plume delineation (a minimum of 6 monitoring wells with at least one un-impacted well)
- Minimum of two years of MTBE data showing a stable or shrinking plume
- Adequate hydrogeologic description (cross-sections, etc.)

Table 2. Site Evaluation

Site Location	Max. Nitrate-N (mg/L)	Sulfate (mg/L)		Methane (μg/L)		Plume Geochemistry	Maximum Hydrogen (nM)	Maximum Ethane (ng/L)
		Min	Max	Min	Max			
Sites recommended for further study								
Millbrae, CA	<0.05	12	120	2400	15000	M+SA	51	1100
Westlake, CA	6.7	240	770	1.7	7100	M+SA	1.7	3200
Redding, CA	<0.05	<5	24	42	3400	M+SD	5.5	2700
Monessen, PA	32	<5	370	4.5	9300	M+SD	1.4	59000
San Mateo, CA	3.5	<5	40	11	2900	M+SD	8.1	2300
San Jose, CA 2	2.9	23	150	1.4	9300	M+SA	2.8	110,000
Agoura Hills, CA	32	14	250	0.08	120	ND+SA	1.0	130
San Jose, CA 1	0.44	280	680	0.23	5.1	ND+SA	2.8	9
Petaluma, CA	5.0	<5	77	5.8	5600	M+SD	6.2	110,000
Belle Meade, NJ	3.2	<5	14	0.29	140	WM+SD	2.0	8
Sites not recommended for further study								
Middletown, NJ	3.1	<5	49	0.06	10,000	M+SD	6.4	2900
Fremont, CA	9.9	29	81	0.17	9700	M+SA	3.5	440
Berkeley, CA 2	2.9	<5	58	1.1	6500	M+SD	21	2800
Lafayette, CA	1.1	14	75	0.15	1800	M+SA	22	230
Mountain View, CA 2	10	8.9	540	2.4	8100	M+SD	5.4	730
Berkeley, CA 1	0.31	<5	42	0.27	6600	M+SD	1.0	<5
Tom's River, NJ	6.8	16	35	0.19	52	ND+SA	2.1	180

Approximately sixty sites were reviewed in light of these criteria and eighteen were selected for Phase II of the API project. This paper presents the results of Phase II.

METHODS

All 18 sites were sampled by the same technician and standard procedures for well purging, sample collection and analysis were used. When possible a peristaltic pump (GeoPump II) was used, for deeper depths a submersible air actuated bladder pump was used (QED ST1102PM). The depth-to-water and total well depth were measured and the pump effluent was passed into a flow-through cell that had a multi-function probe attached to it for readings of dissolved oxygen (DO), pH and conductivity (Horiba U-10). The pump rate was adjusted until the water level in the well remained as close to constant as practical. Approximately one well volume of water was purged and then DO, pH and conductivity readings were recorded every five minutes until three successive readings were within 10% of each other for all three parameters. The well was then considered purged. The pump effluent was attached to a Microseeps hydrogen sampling cell and a dissolved gas sample was collected as per Microseeps' SM9 (McLoughlin and Pirkle, 1999). Sample bottles were then filled for remote laboratory work.

If the DO measured less than 1 mg/L two colorimetric field tests were done. The first test was for dissolved oxygen in the 0.1-1.0 mg/L range (CHEMetrics K-7501). The second test was for ferrous iron in the range 0.5-10 mg/L range (HACH IR-18C). The dissolved oxygen was also measured from the "bubble strip" sample and from a confirmatory water sample. Since dissolved oxygen measurements can be easily corrupted and the bias is almost always high, the minimum of the three measurements was assumed to be the "true" value. The ferrous iron was also measured in the laboratory via Standard Method 3500Fe. Comparison between the results was good qualitatively at all concentrations and quantitatively at concentrations below ~4 mg/L. Ferrous iron is typically used only as a qualitative indicator, so this agreement was all that was required. The field measurement of ferrous iron provided an excellent backup when sample hold-times could not be kept to twenty-four hours. MTBE, TBA, benzene, toluene, all xylenes and the three isomers of tri-methyl benzene (BTEXTMB) were all analyzed via SW846-8260B.

Nitrate and sulfate were measured using an ion chromatograph method, a modified SW846-9056, or via EPA 353.3 for nitrate EPA 375.4 for sulfate. The dissolved gasses were measured via Microseeps' AM20GAX (McLoughlin and Pirkle, 1999).

RESULTS AND DISCUSSION

Plume Geochemistry   Of the 18 sites (103 wells) sampled, reducing conditions were found at most of them, though they varied in strength across each site and the extent of those conditions varied from site to site. All sites had at least some wells that showed a dissolved oxygen level below 1 mg/L. Ferrous iron was detected in at least one sample from all but two sites. Dissolved hydrogen measurements by the "bubble strip" method were possible at 94 of the 103 wells sampled. At the remaining wells, flow could not be sustained long enough for the measurement. A plot of hydrogen concentrations versus ground water contaminant concentration (quantified as a sum total of MTBE, TBA and BTEXTMB concentrations) is presented in Figure 1. Although significant scatter is observed, in general, hydrogen concentration at or above 1 nM were measured in wells with contaminant concentration above 500 µg/L. These hydrogen concentrations are indicative of ongoing sulfate reduction or methanogenesis in the ground water Chapelle, 1997).

At most sites, the maximum concentrations of nitrate-N and sulfate and minimum concentrations of methane represented the geochemistry of un-impacted ground water. Similarly, minimum concentration of sulfate and maximum concentration of methane represented the geochemistry of the impacted ground water. Minimum concentrations of nitrate-N at these sites were below detection limit of 0.05 or 0.1 mg/L (data not presented). Based on the differences in the ground water geochemistry between the impacted and un-impacted wells, the sites were categorized into different geochemical classes as defined in Kolhatkar et al. (2000). Table 1 presents the basis for assigning the plume geochemistry at the sites in Table 2. Going from the bottom to the top of Table 1, the ground water plume becomes more reducing. Based on Table 1 and the conditions observed at the sites, ground water could be classified into one of the following: M+SD, M+SA, WM+SD, WM+SA and ND+SA.

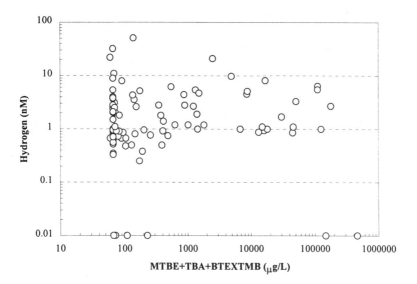

Figure 1. Dissolved hydrogen as a function of contaminant concentration (quantified as a total of MTBE, TBA and BTEXTMB) in ground water

Table 1. Definition of Plume Geochemistry Based on Ground Water Data

Analyte	Concentration in Impacted Wells	Designation
Maximum Methane	> 0.5 mg/L	Methanogenic (M)
	< 0.5 mg/L	Weakly Methanogenic (WM)
Minimum Sulfate	Non-detect or << maximum sulfate	Sulfate depleted (SD)
	Detectable sulfate or < maximum sulfate	Sulfate available (SA)
Minimum Nitrate	Non-detect or << maximum nitrate	Nitrate depleted (ND)

Table 2 summarizes the key ground water data and the corresponding plume geochemistry at these sites. The sites were distributed as follows: methanogenic, sulfate depleted (M+SD, 8 sites), methanogenic, sulfate available (M+SA, 6 sites), weakly methanogenic, sulfate depleted (WM+SD, 1 site) and nitrate depleted, sulfate available (ND+SA, 3 sites). These data indicated that ground water within the plume at 15 out of 18 sites (83%) was significantly reduced (M+SD or M+SA or WM+SD). This was well corroborated by the maximum hydrogen concentrations observed at these sites, which were 1 nM or greater (Table 2) indicative of sulfate reducing or methanogenic conditions. In addition, there was a good correlation between the more reduced ground water conditions (M+SD and M+SA) with elevated maximum ethane concentrations (> 200 ng/L) in ground water. In an earlier ground water survey at BP retail sites (Kolhatkar et al., 2000), ground water at 56 out of 74 sites (76%) was found to be anaerobic (M+SD or M+SA or WM+SD). At some of these sites, a good correlation was observed between methanogenic ground water conditions and strong field evidence of natural MTBE biodegradation. The current study and these past findings underscore the importance of evaluating natural MTBE biodegradation under anaerobic conditions.

Site Selection Criteria for Phase III Work A number of criteria were applied to recommend sites for further work in Phase III of this project.
1. Sites with significant attenuation of MTBE away from the presumed source within the geochemical footprint of the plume were considered for futurre studies (e.g. sites in Milbrae and Petaluma). All wells within the ground water flowpath with dissolved methane concentrations exceeding 0.5 mg/L were considered for this purpose (Kolhatkar et al., 2000). Figure 2 illustrates this behavior at the Milbrae site.
2. Some sites that did not strictly meet the spatial attenuation criterion for this particular sampling event were still considered for Phase III to represent all the

plume geochemistry classes [e.g. sites in San Hose CA 2 (M+SA), Agoura Hills, CA (ND+SA) and Belle Meade, NJ (WM+SD)].
3. Other considerations to recommend sites for further work included a) MTBE only plume at the Milbrae site, and b) significantly high TBA concentrations in comparison to MTBE and BTEX at the Westlake site.

Work proposed in Phase III   Further research work will be conducted at the recommended sites to evaluate the rate and extent of natural MTBE biodegradation under site-specific redox conditions. The detailed scope of this work is being finalized, but will comprise of some or all of the following elements using the soil and ground water samples from the study sites.
1. Static microcosms (with or without ^{14}C-MTBE)
2. Continuous-flow column studies (with or without ^{14}C-MTBE)
3. In-situ microcosm studies at the site
4. Push-pull tests at the site

The experiments will be conducted so as to mimic the site-specific redox conditions.

NATURAL BIODEGRADATION OF MTBE AT A SITE ON LONG ISLAND, NY

Ravi Kolhatkar (BP, Naperville, Illinois)
John Wilson (U.S. EPA, Ada, Oklahoma)
Gordon Hinshalwood (Delta Environmental Consultants, Inc., Armonk, New York)

ABSTRACT : In a plume of gasoline contamination on Long Island, New York, the concentration of MTBE attenuated significantly in certain wells that were down gradient of the source, while the concentration of methane did not attenuate significantly. Near the source, the mingled plume of BTEX and MTBE was uniformly methanogenic. Down gradient of the source, BTEX was completely depleted and the MTBE plume was resolved into two distinct regions, a shallow region with high concentrations of methane (>0.5 mg/liter) and a deeper region with low concentrations of methane. Rates of natural biodegradation were calculated for wells that were in the shallow methanogenic plume. The average ground water seepage velocity at the site is near 700 feet per year. The rate of MTBE attenuation with distance (0.0067 per foot) was estimated as the negative of the slope of an exponential regression of MTBE concentration with distance along the flow path. Using this attenuation rate in the approach of Buscheck and Alcantar (1995), a first order biodegradation rate constant for MTBE was estimated at 5.2 per year (95% confidence interval of 2.2 to 8.2 per year). At this site, MTBE attenuation appears to be strongly correlated to methanogenic conditions in the ground water. Although MTBE is depleted in the shallower methanogenic zone, it persists in a deeper zone in the aquifer where methane is absent.

INTRODUCTION

A case study in North Carolina showed that anaerobic natural biodegradation of MTBE in ground water was associated with methanogenic conditions (Wilson et al., 2000). In order to evaluate whether the occurrence of anaerobic MTBE biodegradation was widespread at spills from gasoline service stations, ground water from seventy-four BP retail sites was surveyed in the summer of 1999 (Kolhatkar et al., 2000). Dissolved methane was used as an indicator of the geochemical footprint of the plume (National Research Council, 2000). At some sites, a first order biodegradation rate constant was calculated from the observed profiles of MTBE concentration versus distance from the source. Only those monitoring wells that had dissolved methane concentrations in excess of 0.5 mg/liter were included in these calculations.

One of these sites is located on Long Island, NY and indicated a relatively rapid rate of MTBE biodegradation. Following the initial survey, additional ground water sampling was conducted at this site in August 2000 and December 2000. This paper presents these groundwater data and interprets the observations.

SITE BACKGROUND

This is a service station site located on Long Island, NY. A gasoline release was first identified in 1990 and the impacted soil was excavated from the source area. Initial ground water remediation efforts at this site included dual-phase extraction using a mobile recovery unit. The site currently has an air sparging/soil vapor extraction (AS/SVE) system in operation. The site is underlain by a sandy aquifer and consists of medium to fine-grained sands with intermixed clay lenses. The depth to ground water is about 20 ft. A hydraulic conductivity of 530 feet/day was estimated based on a 27-hour pumping test conducted in recovery well RW-1 at the northwest corner of the storm water pond (Figure 1). A ground water seepage velocity of 700 feet/year was calculated using an average hydraulic gradient of 0.001 feet/feet and an assumed porosity of 0.3.

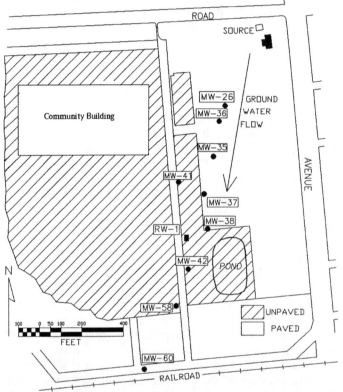

FIGURE 1. Site map showing the locations of monitoring wells and the type of ground surface cover.

The current monitoring network comprises of a number of shallow water table wells at the service station and a number of nested well clusters down gradient. Figure 1 shows the locations of monitoring wells evaluated in this study. The area between the service station and the storm water pond consists primarily of paved parking lots for a grocery store and a strip-mall. A community building is located west of these lots and the ground surface has grass cover. The original source is located below the service station and the wells discussed here are located between 400 ft and 1670 ft from the source area. Figure 2 illustrates the locations of the well screens on these nested well clusters. Each dot represents the location of the center of the screen. Monitoring well 26 is a 2-inch diameter well with a 5-ft long screen just below the water table. Well

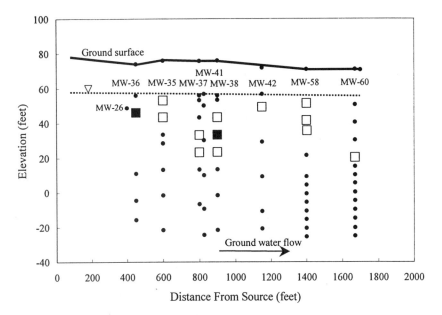

FIGURE 2. Locations of well screens in the cluster wells. Each dot represents the center of the screened interval. Squares represent wells used in the regression shown in FIGURE 7. Closed squares represent wells discussed in FIGURE 3.

locations 36, 35, 37, 41, 38 and 42 have 1-inch diameter vertically nested wells with 5-ft long screens. Additional 1-inch diameter nested well clusters with 1-ft long screens were installed at locations 58 and 60 in September 2000.

MATERIALS AND METHODS

Groundwater samples were collected from well 26 and selected vertical intervals on the wells at locations 36, 35, 37, 41, 38 and 42 during the 1999 survey of BP retail sites. A complete round of groundwater sampling was conducted on the wells at locations 36, 35, 41, 37, 38 and 42 by the USEPA in August 2000. Ground water samples were analyzed for MTBE, TBA, BTEX, sulfate, dissolved methane and a number of biogeochemical parameters. Additional groundwater samples were collected from well locations 58 and 60 in December 2000. Groundwater samples were collected, preserved and analyzed using the methods described previously (Kolhatkar et al., 2000).

RESULTS AND DISCUSSION

Plume Stability Figure 3 illustrates the temporal trends in MTBE (and BTEX) concentrations for a well at location 36 (distance from the source 450 ft) and 38 (900 ft). The particular screened intervals discussed here are shown by closed squares in Figure 2. MTBE concentrations appeared to be relatively stable for both these locations, although some fluctuations were seen, possibly in response to the changes in the water table elevation. These observations suggested that the MTBE plume was stable in this portion of the aquifer. BTEX concentrations also exhibited a similar behavior, but with less fluctuations. BTEX was generally not detected in the wells down gradient of well location 35 (data not presented).

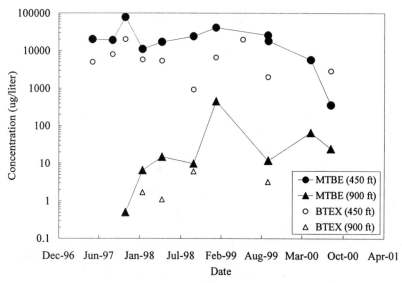

FIGURE 3. Trends in MTBE and BTEX concentrations in wells 450 feet and 900 feet from the source suggesting plume stability.

Fate of MTBE Under Methanogenic Conditions MTBE and dissolved methane concentration contours were plotted using the data collected in August and December 2000 (Figures 4 and 5). A comparison of these plots seems to suggest that MTBE was depleted in methanogenic ground water in the shallower portion of the aquifer (dissolved methane > 0.5 mg/liter). MTBE appeared to persist in the deeper portion of the aquifer where the ground water was not methanogenic.

In comparison to other extensively monitored plumes on Long Island (Haas and Trego, 2000; John Wilson, unpublished data), ground water at this site has become methanogenic in some portions. The surface cover features at this location may explain this. At the service station, the ground water may be aerobic due to the operation of air sparging and SVE systems. However, ground water leaving the source area likely receives minimal recharge from any infiltrating surface water up to well location 38. Most of the soluble electron acceptors (dissolved oxygen, nitrate and sulfate) in the shallow contaminated ground water get depleted due to the natural biodegradation of BTEX and other dissolved organic matter driving it to become methanogenic. However, the deeper portion of the aquifer still seems to contain high sulfate concentrations and this portion has not developed methanogenic conditions (Figure 6).

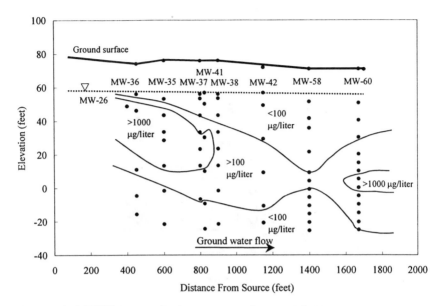

FIGURE 4. MTBE concentration contours in ground water

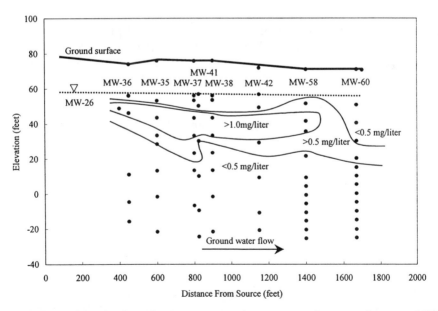

FIGURE 5. Dissolved methane concentration contours in ground water. MTBE is depleted in the shallower methanogenic ground water and persists in the deeper zones where methane is absent.

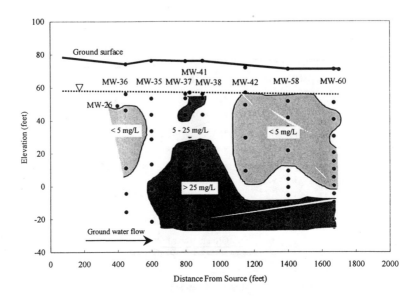

FIGURE 6. Sulfate concentration contours in ground water.

It is noteworthy that the storm water pond located east of well 42 might be acting as a source of recharge on a seasonal basis and may affect the ground water chemistry at locations 42 and down gradient. The MTBE plume appears to have a downward flow component during its travel from well 36 to well 58. This could be explained as a combination of two likely mechanisms; 1) the MTBE plume has followed a downward hydraulic gradient during the seasonal recharge events and, 2) natural biodegradation of MTBE has occurred in the shallower methanogenic ground water and MTBE has persisted in the deeper non-methanogenic portion of the aquifer.

The ground water data in this flow path seem to suggest that MTBE is being naturally biodegraded in the methanogenic ground water. Laboratory microcosm experiments have been set up to further evaluate this possibility. Additional work is underway to characterize the variations in stable carbon isotopic ratios for MTBE, TBA and dissolved methane in ground water at this site.

Biodegradation Rate Constant for MTBE in Methanogenic Ground Water

Between well locations 36 and 35, the mingled plume of BTEX and MTBE was uniformly methanogenic. BTEX was completely depleted further down gradient and the MTBE plume was resolved into two distinct regions, a shallow region with high concentrations of methane (>0.5 mg/liter) and a deeper region with low concentrations of methane. A first order biodegradation rate constant for MTBE was calculated using concentration versus distance data for wells that were in the shallow, stable, methanogenic plume (identified by squares in Figure 2). Figure 7 presents the plots of MTBE and dissolved methane concentrations for these wells within the plume footprint (dissolved methane > 0.5 mg/liter) as a function of distance from the source. These data

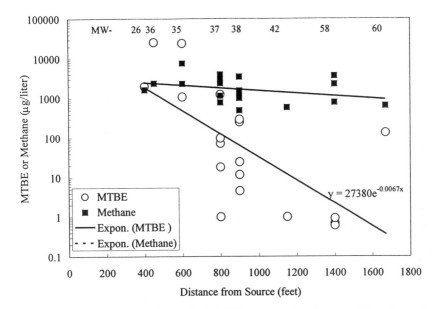

FIGURE 7. Attenuation of MTBE concentrations with distance in the methanogenic ground water (dissolved methane > 0.5 mg/liter). The slope of the regression line for MTBE is statistically different than zero at 95% confidence.

indicate that MTBE is attenuating significantly with distance, whereas, methane remains stable. The rate of MTBE attenuation with distance (0.0067 ± 0.0038 per foot at 95% confidence) was estimated as the negative of the slope of an exponential regression of MTBE concentration with distance along the flow path (Figure 7). Using this attenuation rate in the approach of Buscheck and Alcantar (1995), a first order biodegradation rate constant for MTBE was estimated at 5.2 per year (95% confidence interval of 2.2 to 8.2 per year). A retardation factor of 1.04 was calculated for MTBE based on the following data. The log K_{oc} of MTBE was assumed to be 1.1 (Nichols, 2000). The fraction organic carbon for the soil (f_{oc}) was assumed to be 0.001 (four separate soil analyses varied from 0.00046 to 0.00093). The particle density (ρ) was assumed to be 2.65 gm/cubic centimeter. A longitudinal dispersivity of 27 feet was calculated using the Xu and Eckstein correlation (Xu and Eckstein, 1995).

SUMMARY
1. Significant attenuation of MTBE was observed at this site. This attenuation appeared to be strongly correlated with methanogenic conditions in the ground water. Although MTBE is depleted in the shallower methanogenic zone, it persists in the deeper zone where methane is absent.
2. A first order biodegradation rate constant of 5.2 per year (95% confidence interval of 2.2 to 8.2 per year) was estimated for MTBE in the methanogenic ground water.
3. The ubiquity of the observations made in this study needs to be evaluated at other sites.

ACKNOWLEDGEMENTS

The authors sincerely thank Delta Environmental for their tremendous efforts in coordinating the field activities and compiling the site data. Special thanks are due to Cherri Adair at the R. S. Kerr Environmental Research Center for planning and conducting the August 2000 ground water sampling activities and setting up laboratory microcosms. We acknowledge the excellent analytical support provided by the staff of ManTech Environmental Research Services. Joe Haas with the New York State Department of Environmental Conservation provided useful comments to significantly improve the manuscript.

EPA DISCLAIMER

Although the research described in this manuscript has been supported by the United States Environmental Protection Agency through in-house task 5857 (Natural attenuation of MTBE) at the R. S. Kerr Environmental Research Center, it has not been subjected to Agency review and therefore does not necessarily reflect the views of the Agency and no official endorsement should be inferred.

REFERENCES

Buscheck, T.E., and C.M. Alcantar, 1995. "Regression Techniques and Analytical Solutions to Demonstrate Intrinsic Bioremediation". In Intrinsic Bioremediation, R.E. Hinchee, J.T. Wilson and D.C. Downey, editors. Proceedings of the Third International In Situ and On-Site Bioremediation Symposium, San Diego, California, Volume 3(1): 109-116.

Haas, J. E., D.A. Trego. 2000. "A Field Application of Hydrogen Releasing Compound (HRC) for the Enhanced Bio-Remediation of Methyl Tertiary Butyl Ether (MTBE)". In Proceedings of the 17th Annual Conference on Contaminated soils, Sediments and Water, University of Massachusetts at Amherst, MA, October 16-19, 2000.

Kolhatkar, R., J. Wilson, and L.E. Dunlap. 2000. "Evaluating natural biodegradation of MTBE at multiple UST sites". In Proceedings of the 2000 API/NGWA Conference: Petroleum Hydrocarbons and Organic Chemicals in Groundwater, National Ground Water Association, Westerville, Ohio.

National Research Council, 2000. *Natural Attenuation for Groundwater Remediation*, National Academy Press, page 11.

Nichols, E. M., Beadle, S. C. and Einarson, M. D., 2000. *Strategies for Characterizing Subsurface Releases of Gasoline Containing MTBE*, API Publication Number 4699, A-3.

Wilson J. T., Cho J. S., Wilson, B. H. and Vardy, J. A., 2000. *Natural Attenuation of MTBE in the Subsurface Under Methanogenic Conditions*, EPA/600/R-00/006.

Xu, M. and Y. Eckstein, 1995. "Use of Weighted Least-Squares Method in Evaluation of the Relationship between Dispersivity and Scale". *Ground Water*, 33(6), 905-908.

IMPACT OF NATURAL ATTENUATION AND PHYTOREMEDIATION ON MTBE AND FUEL

Kandi Brown, Larry Tyner, David Grainger, and Tom Perina, IT Corporation, Knoxville, Tennessee; Maureen Leavitt, Newfields, Knoxville, Tennessee; Mike McElligott, Vandenberg AFB, Lompoc, California, USA

Abstract: Groundwater contamination including MTBE and BTEX resulted from leaking fuel storage tanks at the Base Exchange Service Station at Vandenberg AFB, California. The perched groundwater system is hydraulically recharged by a car wash operation and landscaping irrigation system. Although soil excavation was completed during UST removal for the majority of impacted soil, a small volume (estimated 7,800 cubic yards) remained in place, resulting in a continual source of impact to the groundwater. Groundwater balance and flow were highly impacted by existing mature eucalyptus trees surrounding the station. To confirm this impact, a field investigation was completed. Groundwater, rhizosphere soil gas, leaf, and tree core samples were collected. In addition, transpired gas sampling and analysis was completed to determine the potential for contaminant uptake as an indirect consequence of root system water uptake. The impact of natural attenuation was modeled using kriging and linear trend comparisons. The impact of the trees was estimated using the mathematical model PlantX. Linear trends fitted to the data indicated decreases of 12 and 6 pounds per year of BTEX and MTBE respectively. Plant uptake of BTEX and MTBE was estimated between 18 and 20 pounds per year from the groundwater and vadose zone. The data also suggested that phytoremediation had affected the aerial extent of contamination with trees outside the plume boundary containing compounds of concern. Data were used to revise the site fate and transport model for MTBE, reducing the time to achieve cleanup goals from 79 to 13 years. Subsequent confirmation of model predictions indicates BTEX is decreasing at 13 lbs/year and MTBE at 2 lbs/year.

INTRODUCTION

As a result of a leaking UST at the BX Service Station (BXSS), Vandenberg AFB (VAFB), groundwater was impacted at 9 to 28 feet bgs by benzene, toluene, ethyl benzene, and xylene (BTEX) and methyl tertiary butyl ether (MTBE). A 1996 investigation defined the groundwater plume with concentrations of MTBE and benzene ranging up to 15,600 and 31,000 ppb, respectively (Table 1). Approximately 7,800 cubic yards of soil remain as a source; no free product is present. The impacted shallow aquifer is vertically contained by a clay bed (fat clay) dipping from 19 ft bgs to a depth of 28 ft bgs toward the northwest controlling groundwater flow direction. The sand and gravel beds below the deep clay bed are unsaturated and deeper groundwater was not encountered to a depth of 60 feet. There is no evidence of significant water migration through the clay underlying the shallow perched groundwater. The geometric mean hydraulic conductivity of the perched

Table 1
Benzene and MTBE Concentrations in Groundwater Samples - October 2000
BX Service Station, Vandenberg AFB
(Page 1 of 1)

Well	Contaminant	October 1991[a]	February 1993[a]	September 1996[b]	June 1997[b]	November 1997[b]	July 1998[b]	October 1998[b]	May 1999[b]	October 1999[b]	April 2000[b]	October 2000[b]
BXS-MW-1	Benzene	1300	1300	3000	1400	4400	530 UJ	1500 M,J	1000	390	500	560
	MTBE	NA	NA	3600	2700	1200	580 J	540	810	680	350	260
BXS-MW-5	Benzene	7100	31000	6800 T	7300 B,T	7700 T	11000 J	9900 T	13000 U	9400	4800	4100
	MTBE	NA	NA	5500 T	3200 T	3800 T	3800	3900 T	3200 U	3000	3300	1700
BXS-MW-6	Benzene	260	1200	1700	1600 B	2600	3200 J	3200	3200 U	2300	2000	1300
	MTBE	NA	NA	4300	3100	1400	300 J	240	130	65	83	69
MW-14	Benzene	NA	NA	0.5 U	0.5 U	0.5 U	0.4 UJ	3.7	0.4 U	0.4 U	0.31 F	1.60
	MTBE	NA	NA	0.5 U	0.5 U	0.5 U	0.5 U	1.0 U	1.0 U	1.0 U	2.0 U	1.0 U
MW-15	Benzene	NA	NA	0.5 UJ	NA	NA	NA	NA	0.4 U	0.5	0.40 U	0.40 U
	MTBE	NA	NA	0.5 UJ	NA	NA	NA	NA	1.0 U	1 U	1 U	1 U
MW-16	Benzene	NA	NA	0.5 U	NA	NA	NA	NA	0.4 U	0.3	0.40 U	0.40 U
	MTBE	NA	NA	0.5 U	NA	NA	NA	NA	1.0 U	1 U	1 U	1 U
MW-17	Benzene	NA	NA	0.5 U	0.5 U,T	0.5 U,T	0.4 UJ	0.23 F	0.4 U	0.4 U	NA	0.4 U
	MTBE	NA	NA	0.5 U	0.5 UT	0.5 UT	0.5 U	0.50 U	1.0 U	1.0 M	NA	1.0 U
MW-18	Benzene	NA	NA	14.0	NA	NA	NA	7900 J	11000 U	6200	3500	630
	MTBE	NA	NA	80	NA	NA	NA	11000	8100	3400	3200	1300

[a] = Laboratory analysis of benzene was performed by EPA Method 602.
[b] = Laboratory analysis of benzene was performed by EPA Method 8260.
All results reported as micrograms per liter.
BTEX Concentrations reported for 2/92 samples appeared to be reported as mg/L. The concentrations were changed to µg/L.
B - compound present in blank
BTEX - benzene, toluene, ethylbenzene, and total xylenes.
F - Target analyte is detected above the minimum detection limit but below the reporting limit.
J - estimated value
M - A matrix effect was present.
NA - not analyzed
T - Indicates increased turbidity measured >10 NTU during sample collection.
U - detection limit
UJ - The analyte was not detected above reported sample quantitation limit. However, the reported quantitation limit is approximate and may or may not represent the actual limit of quantitation necessary to accurately and precisely measure the analyte in the sample.

zone is 0.53 foot per day (IT, 1997). Because of low permeability and low saturated thickness of the perched zone under the BXSS, the groundwater flow conditions are affected by irrigation, leaking drains, and groundwater uptake by trees.

The eucalyptus trees at the BXSS, planted before the establishment of VAFB, are *Eucalyptus globulus*, the most widely planted species in California (M. McElligot, 1997, personal communication). Eucalyptus trees have high transpiration rates. This trait combined with the solubility and polarity of MTBE and BTEX makes it likely that these compounds are transpired. Trees also facilitate contaminant reduction by: 1) the concentration of an organic in the root of a plant, measured as the root concentration factor, facilitates its removal from the sub-surface environment; 2) the rhizosphere can benefit biodegradation by providing enhanced bacterial populations; however, the presence of organic carbon associated with root exudate may cause inhibition; 3) root exudate can be a beneficial growth substrate for microbes.

Modeling efforts to predict the fate of remaining contamination included the impact of groundwater uptake by existing on-site eucalyptus trees; however, they stopped short of assuming BTEX or MTBE concentrations were reduced by the trees. The goal of a 1998 phytoremediation study was to determine the effects of eucalyptus trees on the BTEX and MTBE concentrations in groundwater and rhizosphere at the BXSS site.

MATERIALS AND METHODS

Groundwater was sampled using low flow sampling techniques and analyzed for volatile and semi-volatile organics according to EPA standard methods.

To measure soil gases, vapor probes were pneumatically driven to within 2 feet of the water table at five points throughout the site using a 1-inch outside diameter steel pipe fitted with a hardened drop-off steel tip (Macho System, KVA, Moshpee, MA). Soil gas samples were analyzed for oxygen, methane, and carbon dioxide using an in-line probe and meter (Landtech Model GA-90 Air Analyzer). To confirm field data, soil vapor samples were collected using a "GAST" diaphragm pump and summa canister.

Transpired gas sampling was carried out on a total of six trees: three trees within, or close to, the impacted area (boundary of the plume) and three trees outside the impacted area (Figure 1). Transpired gas sampling was completed using a method modified from "Draft Demonstration of Phytoremediation of Chlorinated Solvents at Facility 1381, Cape Canaveral Air Station, Florida" (Utah State University and Parsons Engineering Science, Inc., 1997). Glass containers approximately 35 centimeters long and 9 centimeters wide were placed over a secondary limb as large as was practical for the vessel. The bottom of the vessel was sealed gently around the limb, and air was forced through the two side ports. Air was drawn from the vessel with an adjustable air pump at approximately 50 milliliters per minute for 10 hours. As the air exited the vessel, it was filtered through a carbotrap tube to collect organic carbon compounds. All samples were analyzed for benzene, toluene, ethylbenzene, xylenes, MTBE, tert-butyl alcohol, methyl acetate, acetone 2-

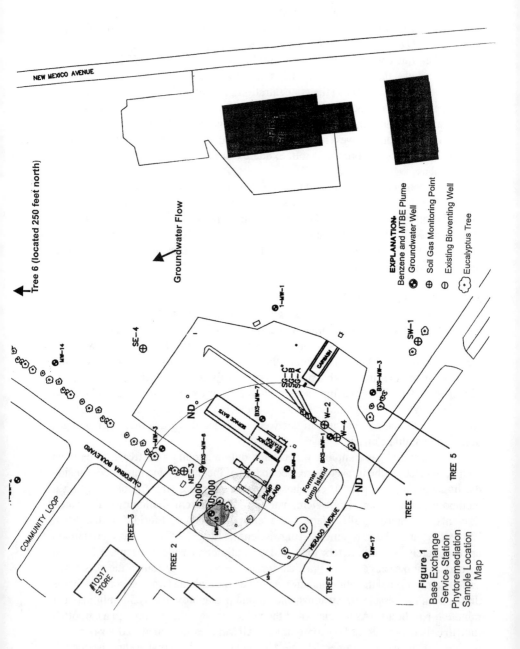

Figure 1 Base Exchange Service Station Phytoremediation Sample Location Map

propanol, methanol and tert-butyl formate (Oregon Graduate Institute). Ambient air samples were collected utilizing the same techniques described above.

Canopy density was derived mathematically through the use of a ceptometer to determine leaf area indices and correlated with subsequent published leaf area indices to correlate transpired gas data to overall transpiration.

Composite core samples were obtained using an 18-inch Haglof Increment Borer (Forestry Suppliers, Inc., Jackson, Mississippi) at approximately 4.5 feet above ground level from areas cleared of outer bark. Two cores were obtained from each tree, from approximately opposite sides of the tree. The tree corer was placed on the trunk surface and turned clockwise until the bit penetrated the surface of the trunk to a depth of approximately 10 inches at which point the extractor was inserted into the corer and a sample was removed. Approximately 7 inches of core material was removed from each boring. A minimum of 30 grams of material was collected and homogenized into a single sample. The samples were analyzed for the same parameters listed above. The hole of the coring was treated with tree wound dressing and sealed with a non-toxic wood putty/sealant. This procedure did not harm the overall welfare of the tree.

Tree root material was obtained using a 2-inch hand-auger approximately 5 feet from the trunk of the tree at a depth of 4 to 5 feet. Tree roots were immediately rinsed with deionized water to remove entrained soil and placed in sample bottles, sealed, labeled, and placed on ice for shipment. Root material was also obtained from within monitoring well MW-18 by the use of a stainless steel 2-inch hand-auger. The hand-auger was lowered into the well, then twisted and pulled upward against the sides of the well casing. The auger bit cut through the root material, which was then trapped between the bit and the side of the casing. Root material obtained was then rinsed with deionized water and containerized. Samples were analyzed for the same parameters listed above.

Samples of leaves were collected from trees within and outside of the impacted area. (Tree 1 and Tree 5, respectively). Samples were collected by stripping numerous secondary branches from within the canopy and placing the leaf material within an 8-ounce jar, sealed, and placed on ice for shipment.

RESULTS AND DISCUSSION

MTBE is generally considered to be resistant to biodegradation. Work at the Vandenberg AFB site provides evidence that MTBE is biodegradable in a subsurface environment. As with many similar natural attenuation projects, the approach to proving that biodegradation is occurring was three-fold. Defining the environmental conditions at the site, determining the presence of degradation byproducts, and extrapolating with a fate and transport model were all used to compile support for this observation.

The environmental conditions on the plume fringe and vadose zone are predominantly oxidative, suggesting that oxygen is the primary electron acceptor. This is an important consideration in determining the potential biodegradation mechanisms. The reports of MTBE biodegradation under oxidative conditions, with aerobic degradation pathways, are more abundant than under anaerobic conditions, although both have been reported in the literature (Hyman and O'Reilly, 1999; Strand, et al., 1999; Hurt, et al., 1999).

The presence of the primary daughter product, TBA, was documented in samples from different groundwater wells at the site as well as in tree samples; however, it is acknowledged that TBA could have been released with the gasoline as a contaminant of MTBE. Generally, the MTBE manufacturing process is approximately 97 percent efficient, suggesting that no more than 3 percent of the MTBE can be present as TBA (Arco Annual Report, 1995). To further explore TBA as a source or a byproduct, the usage of MTBE was considered. MTBE was used to enhance octane in gasoline prior to 1979. For this purpose, MTBE was added at concentrations as high as 8 percent by volume (EPA, 1998). When MTBE was specified as an additive to reduce ozone and smog, recommended usage was between 11 and 15 percent. MTBE-reformulated gasoline has been utilized in California as an ozone reducer. This usage was used as a basis for estimating the original mass of TBA that may have been introduced as a source.

Modeling determined both the historic and current mass of BTEX and MTBE, as well as current estimates of the total mass of TBA. Based on these estimations, significant attenuation of all target constituents has occurred over time. Approximately 90 percent of the BTEX has attenuated, along with 89 percent of the estimated dissolved MTBE. Assuming that BTEX comprises 20 percent of the gasoline present, as much as 445 pounds of gasoline may still be present in the dissolved phase. This suggests that MTBE comprises only 2.5 percent of the total hydrocarbon, significantly less than the EPA directive 11 to 15 percent. Furthermore, the ratio of TBA to MTBE is 14.5 percent, much more than the estimated 3 percent that may have been originally present. The decline in MTBE and the increase in percent TBA suggests that biodegradation may be occurring. This observation is also confirmed when observing concentrations within a given groundwater sample. Where both TBA and MTBE were present, the percent TBA relative to MTBE ranged from 8 to 46 percent.

MTBE and TBA were also found in nearly all of the biomass samples collected. There was no apparent correlation between groundwater concentrations and biomass concentrations. For example, the highest groundwater concentrations were in MW-18, which is adjacent to Tree 2 whose samples contained the lowest concentrations of MTBE and TBA. In another example, Tree 6, estimated to be a downgradient clean tree, showed evidence of the contaminants in its biomass. Researchers have confirmed that unimpacted eucalyptus trees growing under similar climatic conditions (Port Hueneme, California) are nondetect for MTBE and TBA (publication pending, researchers at the University of Washington, 1999). Therefore, the presence of contaminants in Tree 6, outside the present-day plume, could be evidence of plume retraction from its initial areal extent. Only three samples were found to be below the detection limit. Those samples were the core and root samples from Tree 1 (a presumably impacted tree) and the core sample from Tree 4 (on the perimeter of the plume).

All of the biomass samples contained detectable concentrations of BTEX, with a range between 0.532 µg/g in the Tree 2 core sample and 16.9 µg/g in the Tree

3 core sample. The highest concentrations in biomass samples correspond to their proximal location to the plume. Tree 3 is in the immediate vicinity of the plume's leading edge. The roots collected from MW-5 (located in the original source area) contained the second highest BTEX concentration. Tree 5 core sample contained the third highest BTEX concentrations. This may suggest either: a) the tree previously thought to be outside the plume was actually in contact with the contaminants at one time, or b) BTEX may be transported in the vadose zone. Overall, there seems to be a trend of higher concentrations in trees that have come into contact with the plume.

The model of a plant leaf presented within the PlantX model (Dr. Stefan, University of Osnabruck, Germany) was felt to provide an acceptable representation of the dynamics of contaminant transpiration; accordingly, the basic equations of the leaf module of the PlantX model were extracted, and manually utilized to examine the transpiration process. The total mass of the contaminants dissolved in groundwater was calculated by kriging the measured data (using the ordinary kriging algorithm in SURFER; Golden Software, 1994).

The amount of dissolved BTEX and MTBE was shown to be decreasing over time. Linear trends fitted to the data indicate decreases of 13 pounds per year (lbs/year) and 2 lbs/year for BTEX and MTBE, respectively. Plant uptake of BTEX and MTBE was estimated at 17.78 and 19.80 lbs/year. It is important to consider that the trees are removing contamination from the vadose zone as well as from groundwater (the majority of tree roots are above the water table and most of the ground surface near the trees is irrigated daily); therefore, the uptake by trees and decrease in groundwater combined likely represent the overall removal of contamination at the site.

A linear fit was used because of data scatter and short duration of the monitoring data span (about 7 years) relative to the age of the plume (about 29 years).

The results of the phytoremediation investigation indicate that the natural attenuation remedial timing estimate of 79 years, for MTBE was an extremely conservative value. Results of this investigation indicate that this value may be reduced to 13 years once the source is removed.

CONCLUSION

MTBE is biodegraded in the subsurface. This is supported by the observation that TBA is present at levels that far exceed any source material.

The sampling and analytical work using tree samples provide evidence that trees are capable of taking up gasoline-related hydrocarbons. The extension of this information made a significant impact on model predictions for contaminant fate.

While further work should be done to more completely establish this relationship, the direct impact of indigenous trees on organic compounds should be considered in natural attenuation evaluations.

REFERENCES

McElligot, M., Vandenberg AFB, 1997, Personal Communication with M. Leavitt.

Researchers at the University of Washington, 1999, Personal Communication/Publication Pending.

Strand, S. E., L. Newman, P. Heilman, E. Long, and M. Gordon, 1999, *Uptake and Degradiation of MTBE in Trees*, Proceeding from In Situ and On Site Bioremediation, San Diego, California.

Utah State University and Parsons Engineering Science, Inc., 1997, *Draft Demonstrations of Phytoremediation of Chlorinated Solvents at Facility 1381, Cape Canaveral Air Station, Florida*, AFCEE Technology Transfer Division, Brooks Air Force Base, Texas.

NATURAL ATTENUATION OF MTBE IN A DUAL POROSITY AQUIFER

Gary P. Wealthall, Steven F. Thornton and David N. Lerner
(GPRG, University of Sheffield, Sheffield, UK)

ABSTRACT: The natural attenuation of MTBE, TAME and other petroleum hydrocarbons in the UK Chalk aquifer, a fractured dual-porosity formation, is described. Approximately 55,000 L of MTBE and TAME-amended petroleum fuel was accidentally released from underground storage tanks at a retail filling station in early 1999. LNAPL has migrated from the site transverse to the hydraulic gradient and to a depth of 18 m below the water table along the dominant fracture sets. The migration of dissolved contaminants is also controlled by bedding plane fracture dip. Vertical profiles through the aquifer show that contaminant distribution and flux are strongly controlled by fracture transmissivity and recharge events. Groundwater quality data show that MTBE and TAME have migrated further from the site than BTEX. The BTEX compounds are degraded using O_2, NO_3, MnO_2, $Fe(OH)_3$ and SO_4 in the aquifer, whereas the presence of TBA in the groundwater suggests that MTBE is degrading aerobically, downgradient of the BTEX compounds. Mass balances indicate that natural attenuation is effective in reducing the flux of contaminants from this site, by a combination of dilution, sorption, degradation and matrix diffusion.

INTRODUCTION
There is considerable interest in the subsurface fate of methyl tertiary butyl ether (MTBE) and other ether oxygenates that are released following spillages of petroleum fuels amended with these octane enhancers. Compared with BTEX, the oxygenates have higher solubility, lower volatility and lower K_{oc}. This means that spillages of oxygenate-amended fuels will result in the preferential dissolution of the oxygenate into groundwater at higher concentrations, relative to the BTEX constituents. Sorption of the oxygenates to aquifer materials is expected to be lower than for BTEX compounds and the ether oxygenates also appear to be less biodegradable than BTEX. Accordingly, the oxygenates are expected to more mobile, forming large plumes that migrate faster than BTEX. The current understanding of the transport and fate of MTBE in groundwater is generally based on laboratory and field studies in North America. There is very little information on the behaviour of these constituents in UK aquifers (Environment Agency, 2000). The research described in this paper is the first reported study of the natural attenuation of MTBE in a European dual-porosity aquifer.

Site description and aquifer setting. The site is a retail petroleum filling station in southern England and overlies the Chalk aquifer, the most important aquifer system in the UK. The aquifer consists of 99% calcium carbonate ($CaCO_3$), with occasional marl and flint layers but very low mineral oxide (MnO_2, $Fe(OH)_3$) content. The aquifer matrix has high porosity (ca. 30-45%) but low effective permeability, and the matrix pore water is essentially immobile. The matrix is dissected by sets of fractures, which contribute only 1% of the aquifer porosity, but which have very high permeability and dominate groundwater flow in the system. As such the aquifer can be conceptualised as a dual-porosity system, with flow and transport occurring primarily in fractures with high transmissivity and the bulk of groundwater being stored in the low transmissivity porous matrix. In the vicinity of the site, the aquifer is unconfined and overlain by 7 m of drift deposits. The mean water table is ca. 20-22 m below ground level.

Approximately 55,000 L of unleaded fuel was accidentally released from an underground storage tank in February 1999. The spilled fuel contains the ether oxygenates methyl tertiary-butyl ether (MTBE) and tertiary methyl-amyl ether (TAME), at concentrations of 2.88% v/v and 1.65% v/v., respectively. The LNAPL has migrated to the water table, with dissolution of the product resulting in contamination of the saturated zone with a range of petroleum hydrocarbons (diesel range hydrocarbons (DRHC), benzene, toluene, m/p-xylene, o-xylene (BTEX), MTBE, TAME and other aromatic compounds). A soil vapour extraction (SVE) system and groundwater monitoring wells were installed by consultants for the remediation of the source area and evaluation of contaminant migration, respectively.

METHODS
Aquifer rock properties and hydraulic characterisation. Cored boreholes were drilled upgradient of the site and at two locations in the contaminated aquifer. The upgradient borehole was drilled to 55 mbgl, whereas the boreholes in the contaminated aquifer section were drilled to 42 mbgl. Lithological and fracture logs from the rock core samples were compared with downhole Acoustic TeleViewer (ATV) and Optical TeleViewer (OPTV) geophysical logs in the uncased boreholes to identify fracture, style, intensity, orientation, dip and dip direction. Packered pumping tests were used to determine vertical profiles of aquifer transmissivity and contaminant distribution (using onsite chemical analysis) in these boreholes during drilling.

Installation of multilevel samplers and groundwater monitoring. Multilevel samplers (MLS) were installed in each borehole, to monitor groundwater quality at seven horizons identified from the packered pumping tests. Groundwater from the MLS and existing monitoring wells was sampled using dedicated inertial lift pumps. These samples were analysed for petroleum and other volatile/semi-volatile hydrocarbons, MTBE, TAME, tertiary butyl alcohol (TBA), tertiary butyl formate (TBF), total organic carbon (TOC), pH, electrical conductivity (EC), dissolved oxygen (D.O.), Eh, alkalinity, total dissolved inorganic carbon (TDIC), Ca, Mg, Na, K, Fe^{2+}, Mn^{2+}, Cl, NO_3, NO_2, NH_4, SO_4 and S^{2-}, using standard methods. Measurements of groundwater elevation were also made.

RESULTS

Profiles of transmissivity and hydraulic head obtained from the packered pumping tests of MLS boreholes in the contaminated aquifer are shown in Figure 1.

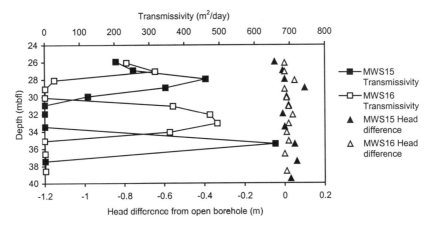

FIGURE 1. Profiles of aquifer transmissivity and hydraulic head

There is no significant difference in hydraulic head between the packer-isolated test zones and the open borehole rest water level in the boreholes located in the contaminated aquifer. This indicates that there is no vertical hydraulic gradient capable of inducing significant vertical migration of contaminants in the aquifer. The transmissivity profiles reveal zones of higher hydraulic conductivity between 27-30 mbfl in the MLS borehole at 30 m from the site (MWS15) and between 26-28 mbfl and 30-35 mbfl in the MLS borehole at 115 m from the site (MWS16).

The distribution of selected species in groundwater 30 m and 115 m from the site is shown in Figure 2 and Figure 3, respectively. Contaminants are present over 20 m depth, between 20-40 m below the filling station (MW7) forecourt level (mbfl) at each location, but with marked spatial heterogeneity in their distribution. Diesel range hydrocarbons (DRHC) and BTEX constituents in the fuel mixture are restricted to a depth of 20-33 mbfl and are below detection limits at 115 m from the site. Contamination with MTBE and TAME extends to a greater depth than with the DRHC or BTEX (Figure 2), but concentrations of the ether compounds are lower in the downgradient MLS (Figure 3).

The concentrations of selected redox-sensitive species in uncontaminated groundwater upgradient of the site are shown in Table 1. The ranges in these species were obtained on two surveys and correspond to the depth interval monitored by the downgradient MLS in the contaminated aquifer. Concentrations of dissolved O_2, NO_3 and SO_4 are zero or below background levels in the contaminated aquifer, whereas concentrations of NO_2, Mn^{2+}, Fe^{2+}, S^{2-} and alkalinity are higher. Values of Eh are close to zero in the zone of BTEX contamination (20-33 mbfl) and are lower at all levels 30m from the site than either background levels or those at 115 m. The SO_4 concentration was similar to background levels and no S^{2-} was detected at 115 m. In general, conditions are anaerobic close to the site at shallow depth in the zone of DRHC and BTEX contamination, but aerobic/sub-oxic at 115 m in the zone of MTBE/TAME contamination. TBA was also detected at µg/l concentrations in groundwater samples from the MLS at 115 m from the site and in some monitoring wells close to the site that are completed with long (10 m) screens.

TABLE 1. Composition of background groundwater, upgradient of the site.

Species	Range	Species	Range
pH	6.83-7.14	SO_4	37-163
Alkalinity (mg/L $CaCO_3$)	240-420	NO_2	bdl-0.8
Eh (mV)	361-416	Mn^{2+}	0.005-0.12
O_2	0.9-4.7	Fe^{2+}	0.01-0.46
NO_3	60-111	S^{2-}	bdl

Notes
All concentrations in mg/L unless stated otherwise; bdl: below detection limits

DISCUSSION AND CONCLUSIONS

The transport and fate of petroleum hydrocarbons in this aquifer is controlled by a combination of aquifer geology, hydrogeology, contaminant properties and groundwater chemistry.

Theoretical calculations suggest that LNAPL has penetrated along vertical fractures to at least 16m below the rest water table. This explains the observed migration of dissolved contaminants to a depth of 40mbfl within 30m distance downgradient of the site, in the absence of a significant vertical hydraulic gradient. Lateral migration of product transverse to the dissolved plume orientation may have occurred along dominant NE-SW to E-W trending high angled fractures. The plume of dissolved phase contaminants is sinking with increasing downgradient transport due to migration along horizontal bedding plane fractures (1-3° dip). A semi-confining hard ground at 64mbgl may limit the depth of plume migration, if this hard ground is laterally continuous.

Groundwater flow occurs primarily along fractures, resulting in the dilution of contaminants with uncontaminated groundwater. This process also occurs by diffusion into the matrix. A mass balance for contaminants at the site suggests that over 90% of the contaminant mass within the saturated zone could be present, dissolved or sorbed within the aquifer matrix. As the source term is exhausted, naturally or by remediation, contaminant concentrations in the fractures will decrease as flux from the site diminishes. Under this condition, contaminants in the matrix will re-diffuse into the fracture flow system, to provide a long term flux to the groundwater and form a secondary source term.

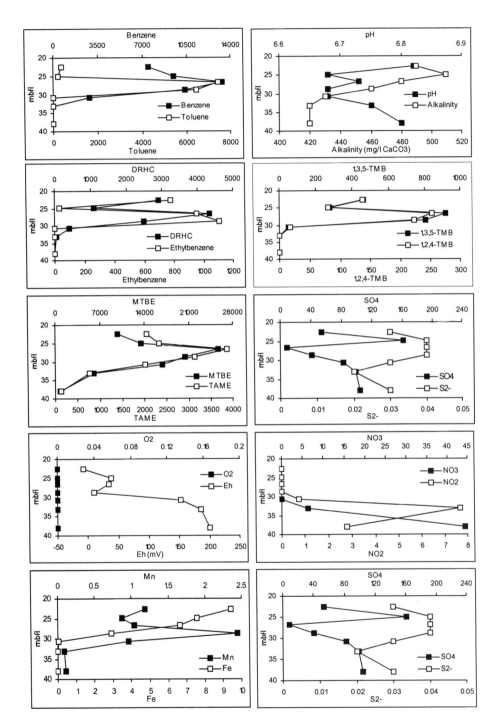

FIGURE 2. Solute concentration-depth profiles at 30 m from the site

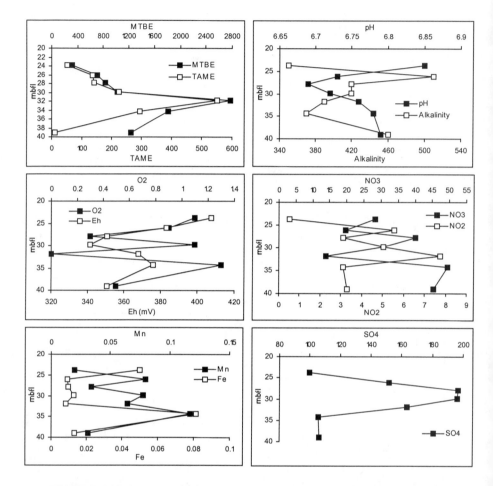

FIGURE 3. Solute concentration-depth profiles at 115 m from the site

Contaminant fluxes from the site and the vertical distribution of contaminants in the aquifer are strongly controlled by water table fluctuations and fracture transmissivity. Contaminant concentrations and fluxes are higher in fractures with higher transmissivity. Episodes of higher contaminant flux from the site are associated with recharge events that increase the water table elevation and allow dissolution of fresh product in fractures that are otherwise not accessible.

Two superimposed plumes are present in the contaminated aquifer. These are a mixed plume of DRHC, BTEX, MTBE and TAME extending from the site to 40 mbfl and 30 m downgradient, and a plume of MTBE and TAME, extending from 40-115 m downgradient of the site to at least 41 mbfl. This distribution reflects the early release of the more soluble oxygenates from the LNAPL. Both MTBE and TAME are transported further from the site, as shown in other studies (Borden et. al., 1997; Landmeyer et. al.; 1998; Schirmer and Barker, 1998).

Degradation of the BTEX compounds using dissolved O_2, NO_3, SO_4 and solid phase MnO_2 and $Fe(OH)_3$ has removed approximately 7% of the dissolved mass in the aquifer. The dissolved oxidants (O_2, NO_3, SO_4) are responsible for most degradation (98%), as the aquifer mineral oxide content is negligible. This degradation estimate represents a minimum based on oxidant flux through the fracture system, and does not consider degradation in the matrix, which may be significantly higher due to the volume of aquifer involved.

Degradation of the BTEX compounds occurs under a wider range of redox conditions, probably reflecting the more biodegradable nature of the BTEX compounds and higher flux of these from the site. The spatial distribution of redox processes, with more anaerobic conditions at shallow depth closer to the site in the zone of higher BTEX concentrations supports this model. Aerobic degradation of MTBE is occurring downgradient of the BTEX compounds, with the production of TBA. A degradation rate cannot be determined for MTBE or TAME with the available data. Based on TBA distribution, MTBE degradation may also occur in the presence of BTEX, closer to the source zone, or at the plume fringe only, where O_2 supply is greater and competition from BTEX for available oxidants is lower. In general, MTBE degradation at this site appears more favourable under aerobic conditions, once the oxygenate plume has migrated downgradient of the BTEX compounds. A conceptual model of natural attenuation for the MTBE and petroleum hydrocarbons at this site is shown in Figure 4. Natural attenuation is effective in reducing the flux of contaminants from this site, by a combination of dilution, sorption, degradation and matrix diffusion.

ACKNOWLEDGEMENTS

The authors gratefully acknowledge the sponsorship of this research by TotalFinaElf UK Ltd. We thank QDS environmental consultants for assistance with field activities at the research site. We also acknowledge the support of CL:AIRE and the Environment Agency in the completion of this work..

REFERENCES

Borden, R.C., Daniel, R.A., LeBrun, L.E. and Davis, C.W. (1997). Intrinsic biodegradation of MTBE and BTEX in a gasoline-contaminated aquifer. *Water Resources Research*, 33, 1105-1115.

Environment Agency (2000). A review of the current MTBE usage and occurrence in groundwater in England and Wales. Environment Agency R&D publication no. 97.

Landmeyer, J.E., Chapelle, F.H., Bradley, P.M., Pankow, J.F., Church, C.D. and Tratnyek, P.G. (1998). Fate of MTBE relative to benzene in a gasoline-contaminated aquifer (1993-98). *Ground Water Monitoring and Remeidiation*, 18, 93-102.

Schirmer, M. and J.F. Barker (1998) A study of long-term MTBE attenuation in the Borden Aquifer, Ontario, Canada. *Ground Water Monitoring and Remediation*, 18, 113-122.

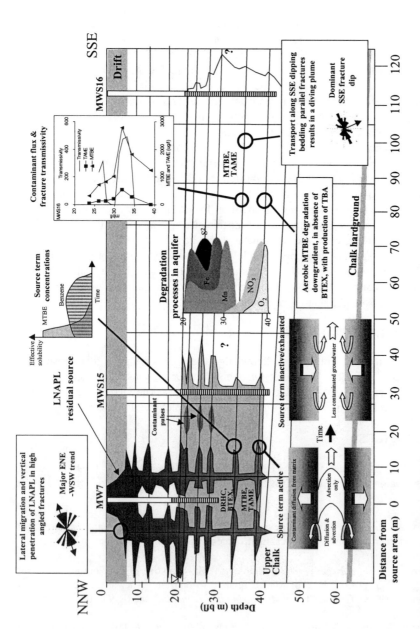

FIGURE 4. Conceptual model for natural attenuation of petroleum hydrocarbons and MTBE in the Chalk aquifer

RELATIVE DEPLETION RATES OF MTBE, BENZENE, AND XYLENE FROM SMEAR ZONE NON-AQUEOUS PHASE LIQUID

Tom R. Peargin (Chevron Research and Technology, Richmond, CA)

ABSTRACT: Smear zone Non-Aqueous Phase Liquid (NAPL) depletion rates of MTBE, benzene, and xylene were studied at 23 UST release sites where groundwater analyses had been performed for several years. Hydrogeologic settings vary, but typically consist of heterogeneous, medium to fine-textured soils. All 23 sites had smear zones containing residual NAPL, and only wells completed and screened in or near the smear zone were selected for the study. A first order decay function was fit to MTBE, benzene, and xylene time series analytical data sets from 15 non-remediation sites and 8 remediation sites to calculate the initial concentration (C_0) and depletion rate (k) for each. Mean k values were similar for all analytes in both site groupings, despite significant differences in predicted relative depletion rates based on solubility and volatility. Unremediated sites yielded k values clustered in a noise band about zero, with both increasing and decreasing trends for all analytes, but at very low rates. Variance from mean k decreased as C_0 increased. Remediation site k values were also similar for all analytes. Mass transfer limitations are commonplace for these hydrogeological settings, and can slow NAPL depletion considerably, effecting MTBE plume shape and length. Mass transfer limitations will also slow the rate of MTBE remediation by source removal technologies relying on dissolution or volatilization.

INTRODUCTION

Based on difference in solubility alone, MTBE should be depleted from smear zone NAPL more rapidly than BTEX. If rapid MTBE depletion from NAPL were commonplace, MTBE plumes derived from UST releases would develop significantly different life cycles than BTEX plumes, including detached MTBE plumes and attached BTEX plumes. Such plumes have been reported in a few sites (Weaver et. al., 1996, Landmeyer, et. al. 1998) which are characterized by coarse-grained soils, high groundwater seepage velocity, and significant recharge. This study examined NAPL depletion rates of MTBE, benzene, and xylene using time series analytical data from 8 UST release sites where remediation has been completed (Peargin, 1999) and 15 sites where remediation has not taken place. Hydrogeologic settings vary, but typically consist of heterogeneous, medium to fine-textured soils. These data provide a unique opportunity to evaluate the degree to which MTBE and BTEX depletion from NAPL source zones are controlled by mass transfer limitations in hydrogeological settings characterized by heterogeneous, primarily fine-grained soil types.

Previous Work. Many workers have performed dissolution experiments to determine partitioning kinetics for NAPL in porous media (Powers et. al., 1992,

1994; Seagren et. al., 1999; Jia et. al., 1999; Garg and Rixey, 1999; Rixey, 2000). Bench-scale models representing homogenous porous media and reasonably low groundwater flow velocities have validated a local equilibrium assumption (LEA) for NAPL partitioning in certain settings, although some workers have noted partitioning at less than equilibrium concentrations where groundwater flow rates are high or NAPL bypassing occurs. Equilibrium partitioning of an individual component of a multi-component NAPL is predicted by Raoult's Law, where dissolved concentration is equal to the product of pure phase solubility and mole fraction in the NAPL mixture. Table 1 lists key physical properties of MTBE, benzene, and xylene affecting NAPL partitioning and depletion rates under conditions of LEA.

TABLE 1. Physical properties of MTBE, benzene, and xylene

Component	MTBE	benzene	xylene
Molecular Weight (g/mole)	88.15	78.11	106.17
Vapor Pressure (atm. @ 20^0C)	0.33	0.1	0.008
Water Solubility (mg/L @ 20^0C)	47,000	1,780	200

Rixey (2000) studied dissolution of MTBE from multi-component NAPL, and found removal rates were consistent with Raoult's Law partitioning, as had also been observed for BTEX in earlier experiments (Garg and Rixey, 1999). Rixey also postulated mass transfer limitations due to NAPL bypassing might slow removal of the final mole fraction of MTBE from NAPL, causing dissolution rates to appear similar to less soluble BTEX compounds over a significant time period.

Durrant et. al. (1999) modeled dissolution and diffusion of MTBE, benzene, and TEX from source NAPL into geologic units of low hydraulic conductivity with subsequent diffusion into the more conductive units once the source area had been depleted and concentration gradients reversed. Their model provides an attractive mechanism for MTBE persistence in the source area for a longer period than would be predicted by advection and dissolution processes alone.

The Air Force Center for Environmental Excellence (AFCEE, 1999) performed a study of BTEX depletion rates from gasoline, JP-4, and JP-5 NAPL through comparison of current NAPL compositions to assumed original compositions and assumed release date. Depletion rates were calculated assuming both zero order and first order partitioning kinetics (termed weathering by AFCEE). Depletion rates were significantly higher than those measured in this study, and are subject to uncertainty through the lack of baseline data.

This study focused on relative MTBE, benzene, and xylene depletion rates from NAPL as evidenced by changes in groundwater dissolved concentrations in monitoring wells completed in, or very near smear zone soils. A smear zone is developed following a UST release as mobile NAPL migrates to the water table and is distributed vertically and laterally through the upper saturated zone. Through seasonal water table fluctuation, the finite NAPL mass is distributed through an increasingly larger soil volume until saturation reduction causes NAPL to become trapped as discontinuous ganglia within soil pores. Modern multi-phase flow theory recognizes the smear zone as an area of intimate contact between NAPL and water, representing a long-term source for dissolved phase contamination (Beckett and Huntley, 1994).

METHOD

Smear zone wells were identified for the 23 study sites by historical presence of measurable liquid hydrocarbon, or high headspace concentrations from soil samples collected at or below the water table. Forty-seven smear zone wells were identified from 15 non-remediation sites and 71 wells were selected from 8 remediation sites. Well sampling was typically performed quarterly, with analysis for MTBE and BTEX by EPA Method 8020 (GC/PID). MTBE was typically included as an analyte throughout the entire monitoring period, or within the first few sampling events.

Regression analysis was used to characterize the rate of groundwater concentration reduction during the monitoring period. Concentration reduction is assumed to be a function of dissolution from NAPL during natural attenuation for the non-remediation sites, and through dissolution and volatilization from NAPL for the remediation sites. A first order decay function was fit to each time series data set using equation 1 (Buscheck and Alcantar, 1995)

$$C_t = C_0 e^{-kt} \quad (1)$$

Where C_t = dissolved phase concentration at time t (ug/L)
C_0 = calculated concentration at time t=0 (ug/L)
k = depletion rate constant (day^{-1})
t = time (days)

Regression using the least-squares method was performed for each time series data set for each analyte. A minimum of 4 groundwater sampling events were required for regression. Non-remediation site time series data sets ranged from 4 to 24 monitoring events, averaging 14 events, while remediation site time series data sets ranged from 4 to 20 monitoring events, and averaged 8 events.

Although groundwater concentration is assumed to directly reflect NAPL composition, additional variability is superimposed by a variety of natural and manmade causes. A screening process was necessary to select data sets in which variance between measured concentrations and the regressed exponential function was low enough for meaningful comparison of depletion rate constants. Standard Error of the Estimate (SeY) was calculated for each data set using equation 2 (Freund, 1984).

$$SeY = \sqrt{\left(\frac{1}{n-2}\right) \Sigma \left(\hat{y}_t - y_t\right)^2} \quad (2)$$

Where y_t = measured concentration at time t (ug/L)
$\hat{y}$ = regressed concentration at correlative time t (ug/L)
n = number of samples in the data set

Since residuals between regressed and measured values increased at low concentrations, values less than 100 ug/L (lower than typical post-remediation source concentrations) were removed from the data sets. An arbitrary maximum variability standard of 0.6 (SeY <0.6) was used to select data sets with acceptably low variability for comparison of k values. Data sets failing screening standards are interpreted to contain excessive natural or manmade "noise" superimposed on natural attenuation or remediation-driven concentration reduction. Thirty-nine remediation site wells and 40 non-remediation site wells passed variance criteria for at least one analyte.

RESULTS

Figures 1 and 2 illustrate characteristic time series data sets for non-remediation and remediation sites, respectively. Non-remediation sites typically show very low to near-zero concentration reduction in smear zone wells, with no significant difference in concentration trends between MTBE, benzene, and xylene.

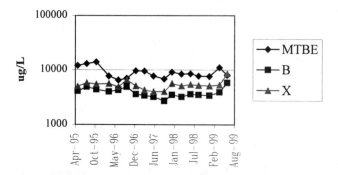

FIGURE 1. Characteristic time series data set for non-remediation site

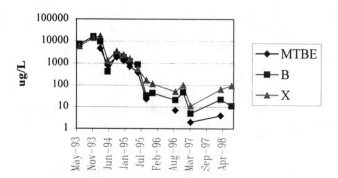

FIGURE 2. Characteristic time series data set for remediation site. Remediation period began April 1994; ended June 1997

Remediation sites show concentration reduction in smear zone wells during the remediation period, but with no significant difference in the rate of concentration reduction between MTBE, benzene, and xylene, despite significant solubility and volatility differences in these compounds.

Figure 3 illustrates xylene time-series regression results plotted as C_0 vs. k for non-remediation sites. Depletion rates (k's) are clustered in a noise band about zero, with time series data sets showing slightly increasing or decreasing trends, but at very low rates. Variation in k's decrease and approach zero as initial concentration (C_0) increases. This is expected as higher initial groundwater concentrations reflect larger NAPL volume in soils surrounding the monitoring well, requiring a larger number of pore volumes (longer time) for NAPL depletion to occur. Benzene and MTBE C_0 vs. k plots show similar patterns.

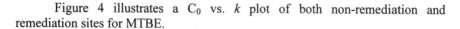

FIGURE 3. Xylene initial concentration (C_0) vs. NAPL depletion rate constant (k) from time-series regression results for non-remediation sites

Figure 4 illustrates a C_0 vs. k plot of both non-remediation and remediation sites for MTBE.

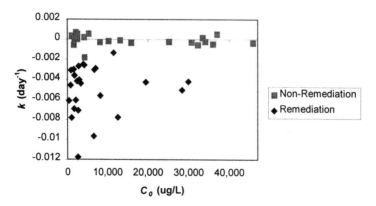

FIGURE 4. MTBE initial concentration (C_0) vs. NAPL depletion rate constant (k) from time-series regression results for non-remediation and remediation sites

Average depletion rates are approximately an order of magnitude greater for remediation sites as compared to non-remediation sites. Remediation site C_0 vs. k plot patterns are similar for benzene, and xylene.

Tables 2 and 3 list key regression analysis statistics of time series data sets from non-remediation and remediation sites, respectively. These data demonstrate there is little statistical difference between mean NAPL depletion rates for MTBE, benzene, and xylene for each site group.

TABLE 2. Regression analysis statistics of time series data sets from non-remediation sites

Parameter	MTBE	Benzene	Xylene
Time-series data sets	22	39	34
Mean NAPL depletion rate constant (k)	-1.1E-04	-3.7E-04	-2.0E-04
Standard deviation of k set	5.3E-04	6.7E-04	7.8E-04

TABLE 3. Regression analysis statistics of time series data sets from remediation sites

Parameter	MTBE	benzene	Xylene
Time-series data sets	34	30	37
Mean NAPL depletion rate constant (k)	-3.5E-03	-2.2E-03	-3.0E-03
Standard deviation of k set	3.0E-03	2.1E-03	2.0E-03

CONCLUSIONS

Regression analysis of MTBE, benzene, and xylene time series data sets from 23 sites underlain by heterogeneous, medium to fine-textured soils illustrates mass transfer limitations affect NAPL depletion rates under both remediation, and non-remediation (natural attenuation) conditions. NAPL bypassing mechanisms described by Rixey (2000) producing apparently similar depletion rates for MTBE as compared to other BTEX compounds may apply to non-remediation sites, but are hard to reconcile with the large number of pore volume exchanges occurring during remediation. Diffusion-limited mass transfer described by Durrant et. al. (1999) may better account for observed similarity between depletion rates for these analytes as their aqueous diffusion coefficients are essentially identical, and low conductivity soils were present at all sites studied.

Regardless of mechanism, these data suggest NAPL mass transfer limitations are commonplace, and can slow the overall source depletion process considerably. Mass transfer limitations should effect observed MTBE plume shape and length under conditions of natural attenuation, and probably account for the limited number of documented detached MTBE plumes. Mass transfer limitations will also slow the rate of MTBE removal during remediation with technologies relying on dissolution or volatilization of source zone NAPL.

REFERENCES

Air Force Center for Environmental Excellence. 1999. *Light Non-Aqueous Phase Liquid Weathering at Various Fuel Release Sites.* F41624-92-D-8036 no.25. 93 p.

Beckett, G.D., and D. J. Huntley. 1994. "The Effect of Soil Characteristics on Free-Phase Hydrocarbon Recovery Rates." *Proc. Petroleum Hydrocarbons and Organic Chemicals in Groundwater: Prevention, Detection, and Restoration.* National Groundwater Association, Houston, TX. November 2-4. pp. 511-525.

Brusseau, M.L. 1992. "Rate-Limited Mass Transfer and Transport of Organic Solutes in Porous Media that Contain Immobile Immiscible Organic Liquid." *Water Resources Research.* 28(1) 33-45.

Buscheck, T.E., and Alcantar, C.M. 1995. "Regression Techniques and Analytical Solutions to Demonstrate Intrinsic Bioremediation." In R.E. Hinchee, J.T. Wilson, and D.C. Downey (Eds.), *Intrinsic Bioremediation.* pp. 109-116. Battelle Press, Columbus, OH.

Durrant, G.C., Schirmer, M., Einarson, M.D., Wilson, R.D., and Mackay D.M., 1999. "Assessment of the Dissolution of Gasoline Containing MTBE at LUST Site 60, Vandenberg Air Force Base, California." *Proc. Petroleum Hydrocarbons and Organic Chemicals in Groundwater: Prevention, Detection, and Restoration.* National Groundwater Association, Houston, TX. November 17-19. pp. 158-166.

Fruend, J.E. 1984. *Modern Elementary Statistics.* Prentice-Hall Inc. Englewood Cliffs, NJ

Garg, S. and Rixey, W.G. 1999. "The Dissolution Of Benzene, Toluene, M-Xylene And Naphthalene From A Residually Trapped Non-Aqueous Phase Liquid Under Mass Transfer Limited Conditions." *Journal of Contaminant Hydrology.* 36(1999) pp. 313-331.

Jia, C., Shing, K., and Yortsos, Y.C. 1999. "Visualization and simulation of non-aqueous phase liquids solubilization in pore networks." *Journal of Contaminant Hydrology.* 35(1999) pp. 363-387.

Landmeyer, J.E., Chapelle, F.H., Bradley, P.M., Pankow, J.F., Church, C.D., and Tratnyek, P.G. 1998. "Fate of MTBE relative to benzene in a gasoline-contaminated aquifer (1993-98)." *Ground Water Monitoring Remediation.* Fall: pp. 93-102

Peargin, T.R., 1999. "An Empirical Study of MTBE, Benzene, and Xylene Groundwater Remediation Rates." In Alleman, B.C., and Leeson, A. (Eds.) *In*

Situ Bioremediation of Petroleum Hydrocarbon and Other Organic Compounds. Battelle Press, Columbus, OH.

Powers, S.E., Abriola, L.M., and Weber, W.J.Jr. 1992. "An Experimental Investigation of Non-aqueous Phase Liquid Dissolution In Saturated Subsurface Systems: Steady State Mass Transfer Limitations." *Water Resources Research* 28(10) pp. 2691-2705.

Powers, S.E., Abriola, L.M., and Weber, W.J.Jr., 1994. "An Experimental Investigation of Non-aqueous Phase Liquid Dissolution in Saturated Subsurface Systems: Transient Mass Transfer Limitations." *Water Resources Research* 30(2) pp. 321-332.

Rixey, W.G., 2000. "Dissolution of MTBE from a Residually Trapped Gasoline Source." *American Petroleum Institute Technical Research Bulletin.* July 20, 11p.

Seagren, E.A., Rittmann, B.E., and Valocchi, A.J. 1999. "A Critical Evaluation of the Local-Equilibrium Assumption in Modeling NAPL-Pool Dissolution." *Journal of Contaminant Hydrology.* 39(1999) pp. 109-135.

Weaver, J.W, J.T. Haas, and J.T. Wilson. 1996. "Analysis of the gasoline spill at East Patchogue, New York." *Procedings of Non-Aqueous Phase Liquids (NAPLs) in Subsurface Environment: Assessment and Remeidation* L. Reddi (Ed.) American Society of Civil Engineers, Washington, D.C., November 12-14, pp. 707-718.

NATURAL ATTENUATION OF DISSOLVED BENZENE AND MTBE - TWO CASE STUDIES

Joseph Robb and Ellen Moyer (ENSR International, Westford, Massachusetts, USA)

ABSTRACT: This paper compares the natural attenuation of benzene and methyl tertiary butyl ether (MTBE) at two Midwestern U.S. retail gasoline marketing outlets. The two sites were chosen to compare a site where natural attenuation clearly could play a role in site management, with a site where concentration trends indicate additional engineered remediation may be necessary. Dissolved plumes of benzene and MTBE were characterized as stable, expanding or shrinking based on non-parametric Mann-Kendall trend evaluations. In order to demonstrate stable or decreasing contaminant plumes, time trend evaluations were performed to ascertain the behavior of benzene and MTBE at individual monitoring wells. Trend evaluation results from individual monitoring wells were then interpreted within the hydrogeologic context of each site to describe site-wide trends in plume behavior. Datasets that did not exhibit upward or downward trends were further characterized as stable or non-stable trends with the coefficient of variation (CV) test. The attenuation rates were quantified using Sen's non-parametric indicator of median slope. Dissolved oxygen (DO) and oxidation-reduction potential (ORP) data were evaluated to qualitatively assess the potential for biological degradation of benzene and MTBE. Geochemical parameters indicate biological degradation is occurring at both sites. Significant downward trends in benzene and MTBE suggest natural attenuation could play a role in site management at one site, while steady and increasing trends in benzene and MTBE suggest source control measures may be needed at the other site.

INTRODUCTION

Objectives. The objectives of this paper are to:

- Assess the natural attenuation of benzene and MTBE using datasets "typical" of gasoline station sites (i.e., data from a limited monitoring well network that are not collected with the primary goal of evaluating natural attenuation).
- Use the Mann-Kendall trend test and the coefficient of variation (CV) test to identify decreasing, stable or increasing concentration trends at individual wells and, by extension, identify decreasing, stable or increasing plumes.
- Use Sen's nonparametric trend estimator to quantify the magnitude of concentration trends.
- Demonstrate how trend evaluations can also be used to help identify the need for additional source control measures or site characterization work.

Trend Analysis. The non-parametric Mann-Kendall test and Sen's non-parametric trend estimator were combined with the CV test to evaluate time trends in benzene and MTBE data. The Mann-Kendall test and Sen's trend estimator are considered well suited to the dataset because they can be used on data that are non-parametric (do not have a specific distribution, such as normal or log normal) and the dataset can contain data at irregularly spaced intervals. In addition, the dataset can contain elevated values compared to the average (outliers) or data reported as below the practical quantitation limit. The following guidelines were followed when applying and interpreting the results from the Mann-Kendall test and Sen's estimate of trend:

- An α-significance level of 90% was deemed appropriate for natural attenuation trend evaluations.
- To remove the bias introduced by detection limits that change over time, all non-detect values were assigned one half of the lowest detection limit.
- Mann-Kendall evaluations were not performed for time series data that consisted entirely of points near and below the detection limit, due to the inherent uncertainty of analytical data near the detection limit.
- Mann-Kendall trend evaluations were not performed on datasets that exhibited seasonality.
- Sen's non-parametric estimator of median slope was used to quantify downward trends in benzene and MTBE at individual monitoring wells.
- Trend results for individual wells were interpreted considering the location of source area(s), groundwater flow directions and approximate dissolved plume locations.

Datasets that did not exhibit upward or downward trends with the Mann-Kendall analysis were further characterized as stable or non-stable trends with the coefficient of variation (CV) test. The CV, calculated by dividing the standard deviation by the arithmetic mean, was used to measure the variability within each time series dataset. The CV should be less than or equal to 1 for a Mann-Kendall "no upward or downward trend" result to be considered a stable trend. If the CV is greater than 1, the trend is considered unstable.

Geochemical Data. It is now recognized that biological degradation of benzene and MTBE can occur under a variety of terminal electron accepting processes. Although the data were insufficient to precisely identify terminal electron accepting processes, general observations of trends in redox chemistry across the site were made. In general, lower ORP and DO values in the source area, when compared to background, were considered good evidence of the consumption of oxygen for the biological degradation of benzene and MTBE.

SITE A.

Site A was reported to regulatory authorities in January 1992 when light non-aqueous phase liquid (LNAPL) was identified in the vicinity of the former dispensing island, canopy footings and dispenser lines during excavation activities associated with UST upgrades. An unspecified amount of hydrocarbon-impacted soil was removed during the UST upgrades, but no further engineered soil or groundwater remediation has been performed at Site A. No LNAPL has been detected in site monitoring wells. See Figure 1 for locations of site features.

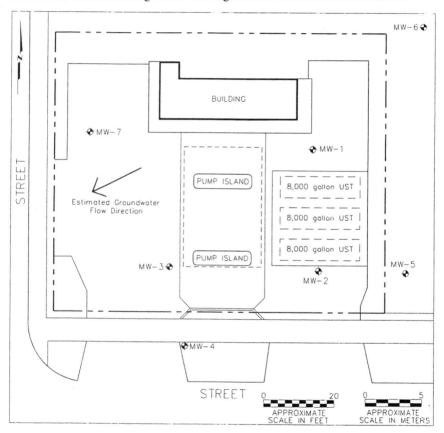

FIGURE 1. Plan view of Site A.

Hydrogeology. The depth to groundwater at Site A is 5 to 12 feet (1.5 to 3.7 meters) below ground surface (bgs). The shallow unconfined aquifer consists of unconsolidated glacial deposits that are reddish brown medium stiff clay, containing a trace of silt and fine sand. The horizontal groundwater flow direction is to the southwest and the average horizontal groundwater velocity is estimated to be approximately 110 feet per year (33.5 meters per year).

Seasonality. Between January 1995 and July 1998, increases in groundwater elevations generally coincided with increases in dissolved benzene concentrations in source area monitoring wells MW-1, MW-2, MW-3 and MW-4. Since Site A is paved and relatively impervious to infiltration, increases in dissolved benzene concentrations could be due to the seasonal rise of the water table into zones of residual soil contamination. After January 1998, seasonal effects appear to play a smaller role in concentration trends, which indicates a potential depletion of residual soil contamination within the range of seasonal groundwater fluctuations. Figure 2 presents groundwater elevations and the concentrations of total benzene and MTBE at monitoring well MW-1 for the period from January 1995 to November 2000. Similar trends were observed at other source area monitoring wells.

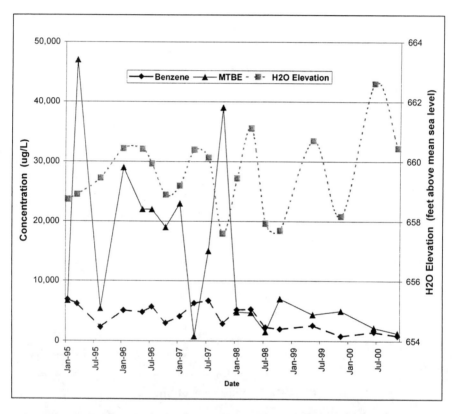

FIGURE 2. Concentration trends and groundwater elevations at MW-1 at Site A.

Trends. Mann-Kendall trend analyses for MTBE were evaluated over ten sampling rounds from July 1997 to November 2000 and over seven sampling rounds from April 1998 to November 2000 for benzene. Mann-Kendall trend analyses were not performed for benzene data collected prior to April 1998, or for MTBE data prior to July 1997, due to the presence of seasonal trends (see Figure 2). Mann-Kendall trend analyses indicate decreasing trends in benzene and MTBE concentrations at the 90% confidence level in source area monitoring wells MW-1, MW-2, and MW-3, as shown in Table 1. Source area monitoring well MW-4 exhibits a stable trend in MTBE and an unstable trend in benzene concentrations.

TABLE 1. Site A source area benzene and MTBE trend analysis.

Date	MW-1 MTBE (µg/L)	MW-1 Benzene (µg/L)	MW-2 MTBE (µg/L)	MW-2 Benzene (µg/L)	MW-3 MTBE (µg/L)	MW-3 Benzene (µg/L)	MW-4 MTBE (µg/L)	MW-4 Benzene (µg/L)
7/97	15,000		4,500		190		460	
11/97	39,000		20,000		400		670	
1/98	4,800		50		2,800		2	
4/98	4,700	5,300	2,900	3,400	500	1,400	520	270
7/98	1,500	2,300	2,100	3,100	590	1,900	180	15
11/98	7,000	2,000	2,900	3,100	300	1,300	140	4.3
5/99	4,400	2,600	3,300	2,800	140	760	110	48
11/99	5,000	820	2,700	1,800	13	780	190	6
5/00	2,200	1,500	790	1,400	96	670	62	20
11/00	1,300	850	920	2,100	70	680	150	7
Decreasing Trend?	Yes	Yes	Yes	Yes	Yes	Yes	No CV < 1	No CV > 1
Slope (µg/L per year)	-2024	-620	-682	-516	-222	-377	-40	1

Decreasing concentrations of benzene and MTBE in wells MW-1, MW-2, and MW-3, are strong evidence of an attenuating source of benzene and MTBE. Where benzene concentrations are highest, (MW-1, MW-2 and MW-3), Sen's slope estimates show benzene is decreasing at rates between 377 µg/L and 620 µg/L per year. At these same wells, MTBE is decreasing at rates between 222 µg/L and 2,024 µg/L per year. The highest estimate of MTBE decline (2,024 µg/L per year at MW-1) may be an overestimate due to a single elevated concentration value in October 1997 and lower values in subsequent sampling rounds. In general, the slope estimates suggest benzene and MTBE concentrations are declining at approximately the same rate in source area wells.

Trend evaluations were not performed for volatile organic compound (VOC) data from perimeter wells MW-5, MW-6 and MW-7 because the data consisted primarily of concentration values near or below the detection limit (see discussion above). These wells effectively delineate the extent of the dissolved VOC plume to the northeast (hydraulically upgradient) and northwest (cross-gradient).

Geochemical Conditions. Table 2 provides April 1998 to November 2000 average MTBE, benzene, DO and ORP data.

Table 2. Average MTBE, benzene, DO and ORP at Site A.

Well ID	MTBE (µg/L)	Benzene (µg/L)	Dissolved Oxygen (mg/L)	Oxidation-Reduction Potential (mV)
MW-1	3,729	2,196	3.3	-33
MW-2	2,230	2,539	4.3	-48
MW-3	283	1,070	4.3	-40
MW-4	193	53	4.3	-27
MW-5	1.3	< 1	5.0	80
MW-6	1.4	< 1	3.7	99
MW-7	< 1	< 1	4.5	62

In general, there is a good correlation between the presence of MTBE and benzene, and low ORP values, relative to background values. In addition, the DO concentration is higher in background well MW-5, and slightly lower in the source area, cross-gradient and downgradient wells. These results suggest DO is being used as a terminal electron acceptor in the source area.

SITE B

Site B was reported to regulatory authorities in August 1993 when photoionization detector field screening identified petroleum-impacted soil in the vicinity of the gasoline transfer lines. An unspecified amount of soil was removed but no LNAPL was observed during the 1993 excavations. No LNAPL has been observed in Site B monitoring wells. See Figure 3 for locations of site features.

The shallow unconfined aquifer consists of unconsolidated glacial deposits that are dark gray silt with some clay and a trace of fine gravel and coarse sand. The average horizontal groundwater velocity is estimated to range from 0.2 to 20 feet per year (0.06 to 6 meters per year). The depth to groundwater ranges from 3 to 11 feet (0.9 to 3.3 meters) bgs, and the horizontal groundwater flow direction is to the southeast. The distribution of contaminants at the site is consistent with the timing of the release and the estimated groundwater velocity.

Table 3 describes concentrations of benzene and MTBE detected historically at each location, statistical trend evaluation results and average DO and ORP data. Approximately 20 rounds of data were available from April 1994 to November 2000. Data prior to April 1997 appeared to exhibit seasonal trends and were therefore excluded from the trend evaluations.

The elevated, stable benzene and MTBE concentrations at MW-2, located immediately downgradient of the gasoline-dispensing island, provide strong evidence of a continuing source of petroleum hydrocarbons. The stable trend in MTBE at well MW-3 and upward trend at MW-1 indicate the plume of MTBE may be stable in the northeast, but is expanding in the western part of the site. Elevated MTBE and benzene concentrations at well MW-2 indicate dissolved benzene and MTBE are migrating off site.

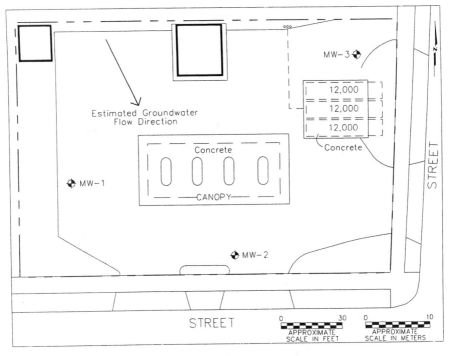

FIGURE 3. Plan view of Site B.

TABLE 3. Site B Contaminant distribution, benzene and MTBE trends and geochemical parameters.

Well ID	April 1997 to November 2000 Benzene and MTBE Concentrations (10 sampling rounds)	Trend Evaluation Results	Average DO (mg/L)	Average ORP (mV)
MW-1	No benzene detected	Benzene not analyzed (see discussion above)	4.3	78
	MTBE ranges from 10 to 85 µg/L	MTBE exhibits upward trend		
MW-2	Benzene 400 to 3,000 µg/L	Benzene exhibits stable trend, CV<1	3.3	-18
	MTBE 68 to 5,000 µg/L	MTBE exhibits stable trend, CV<1		
MW-3	Benzene ranges from non-detect (ND) to 10 µg/L	Benzene exhibits unstable trend, CV > 1	4.5	60
	MTBE ranges from 100 to 1,000 µg/L	MTBE exhibits stable trend, CV < 1		

Average DO and ORP values are lowest where benzene and MTBE concentrations are highest (MW-2), and higher average ORP and DO values were measured at wells MW-1 and MW-3, which contain little or no benzene and low to moderate concentrations of MTBE. The DO, ORP, benzene and MTBE results are consistent with the consumption of oxygen for the biological degradation of gasoline constituents.

Overall, results indicate biodegradation may be limiting the migration of benzene and MTBE. However, residual contamination is acting as a continuing source, the extent of dissolved benzene appears to have reached steady state and the dissolved plume of MTBE may be expanding. Additional source control measures may be warranted.

CONCLUSIONS

Two petroleum retail marketing outlets with documented releases of petroleum hydrocarbons were evaluated to characterize the natural attenuation of benzene and MTBE. Dissolved plumes were characterized as stable, expanding or shrinking based on the non-parametric Mann-Kendall trend evaluations. Datasets that did not exhibit upward or downward trends were further characterized as stable or non-stable trends with the coefficient of variation test. The slopes of downward trends were quantified using Sen's non-parametric indicator of median slope. DO and ORP values were reviewed to qualitatively assess the contribution of biodegradation.

In general, average DO and ORP values were lowest where benzene and MTBE concentrations were highest, and higher average ORP and DO values were measured at locations with little or no benzene and low to moderate concentrations of MTBE. The DO, ORP, benzene and MTBE results at Site A and Site B are consistent with the consumption of oxygen for the biological degradation of gasoline constituents and suggest biodegradation contributes to the overall natural attenuation of petroleum hydrocarbons at both sites.

The trend evaluations provided a preliminary evaluation of the potential for natural attenuation to play a role in management of each of the sites. Trend evaluations at Site A provided good evidence of a decreasing source and shrinking or stable benzene and MTBE plumes. Based on these preliminary results, natural attenuation could clearly play a role in the management of Site A. Site B showed evidence of a continuing source, a steady state dissolved plume of benzene and a potentially expanding plume of MTBE. At Site B, natural attenuation may be a viable management tool in combination with additional source control measures.

HYDROLYSIS OF MTBE: IMPLICATIONS FOR ANAEROBIC AND ABIOTIC NATURAL ATTENUATION

Kirk T. O'Reilly[1], Michael E. Moir[1], Christine Taylor[2], and Michael R. Hyman[2]
[1] Chevron Research and Technology Co., Richmond, CA, 94803
[2] North Carolina State University, Raleigh, NC 27695

ABSTRACT: Methyl *tertiary*-Butyl Ether (MTBE) is generally considered to be resistant to chemical transformation in the aqueous phase. This lack of reactivity has lead to concerns of the long term impacts of MTBE in groundwater. Although hydrolysis has been recognized as a mechanism for MTBE transformation, it has been discounted as a significant reaction under environmental conditions. This idea is reinforced by the chemical literature that indicates alkyl ethers are only sensitive to hydrolysis under strongly acidic conditions. In this study, we have examined the fate of MTBE and other ether oxygenates under moderately acidic conditions ($\geq$ pH 2). Our results demonstrate that MTBE is sensitive to acid-catalyzed hydrolysis reaction that generates *tertiary* butyl alcohol and methanol as products. The transformation rate decreases with pH and increases with temperature. A rate of 1×10^{-3} d^{-1} was measured at pH 2 and 25 °C. The results will be discussed in terms of the potential implication for the natural attenuation of MTBE. A hypothesis of abiotic soil catalyzed hydrolysis will be presented. Hydrolysis also has implications for microbial degradation. Although oxygenase enzymes are typically responsible for MTBE transformation in aerobic organisms, these findings suggest a potential anaerobic enzymatic mechanism.

INTRODUCTION:

Methyl *tert*-Butyl Ether (MTBE), a commonly used gasoline oxygenate, is generally considered to be resistant to chemical transformation in the aqueous phase (1,2). This lack of chemical reactivity has led to concerns over of the long-term impacts of MTBE contamination in groundwater. Although a review of the chemical literature (3) clearly indicates that alkyl ethers are typically sensitive to hydrolysis under strongly acidic conditions, acid-catalyzed hydrolysis of MTBE has been discounted as a mechanism for MTBE transformation under environmental conditions (4). The study of ether cleavage and hydrolysis has a long and important history spanning nearly 150 years of the chemical literature. The first observation of ether cleavage was made in 1861, the reaction taking place in concentrated hydroiodic acid (5). The first report of the cleavage of alkyl ethers was published by Silva in 1875, work also performed in concentrated hydroiodic acid (6). In 1893, the first systematic study of the effect of ether structure on reaction rate was performed (7). In another study, a tertiary ether was cleaved by aqueous hydroiodic acid in 15 minutes at room temperature (8). The first observation of MTBE cleavage was made in 1932, where the reaction was observed to proceed in a matter of hours at 0° C in 27% HCl (9). Church et al. (10) discussed the potential for the hydrolysis of MTBE and its degradative intermediate *tertiary* butyl formate under environmental conditions.

We recently presented findings of our investigation of the hydrolysis of MTBE and other ether oxygenates (11). The study was initiated when we unexpectedly observed high concentrations of TBA in experiments where the bacterial transformation of MTBE had been halted by acid-quenching. The results demonstrate that ether oxygenates are susceptible to acid hydrolysis at moderate pH values. In this paper these findings are discussed in terms of their possible impacts on our understanding of the fate of MTBE in the ground water environment.

METHODS

MTBE (99.9%) was obtained from either Burdick and Jackson (Muskegon, MI) or Aldrich Chemical Co., (Milwaukee, WI). *Tertiary* amyl methyl ether (TAME) (97%), ethyl *tertiary* butyl ether (ETBE) (99%), *tertiary* amyl alcohol (TAA) (99%) and *tertiary* butyl alcohol (TBA) (99.5%) were obtained from Aldrich Chemical Co.

The pH of distilled water was adjusted with the experimental acid or buffer and MTBE was added on a weight basis to the desired initial concentration. The solution was transferred to 40 mL volatile organic analysis (VOA) vials (VWR Scientific, Boston, MA)) that were completely filled to exclude headspace. The vials were then sealed with PTFE-lined septa. Sufficient vials were prepared to allow two to be sampled at each sampling time and each vial was only sampled once. The vials were kept in the dark at 25 °C without shaking. The vials were sampled by removing the lid, and pipeting 2 ml of the reaction mixture into gas chromatography (GC) autosampler vials. 1 µL samples of the solution were analyzed for the concentration of MTBE, TBA and methanol by splitless injection into a Hewlett-Packard 6890 GC (Wilminton, DE) equipped with a 15m x 0.32 J&W DB-1 column (Folsom, CA) and FID detector. Quantification of MTBE and TBA was performed using external standard calibration.

Most experiments were run for 4 days. Samples were analyzed daily. Reaction rates were calculated based on the measured increase in TBA concentration as a function of MTBE concentration over time.

The influence of pH on the rate of MTBE hydrolysis was determined in incubations in which the pH of sodium phosphate buffer (0.2 M) was adjusted to pH values of 2. 2.5, 3, 4, 5, 6, or 7 using concentrated HCl (~12 N). The hydrolysis of MTBE in this sodium phosphate-buffered solution was also evaluated at pH 2 using four initial concentrations of MTBE (10, 25, 50 and 100 mM) and at two different temperatures (25 and 37° C). The hydrolysis of TAME and ETBE (both at 100 mM) was also evaluated at pH values of 2 and 7 in incubations conducted using sodium phosphate-buffered solutions.

In O'Reilly et al. (11), the ability of strong acid ion exchange resins to catalyze the hydrolysis of MTBE was demonstrated. In the current study, two naturally acidic solid materials were tested for their ability to catalyze the hydrolysis of MTBE. One was peat moss (Lakeland Peat Moss LTD., Edmonton, Canada) and the other was Montmorillonite clay (Ward's Natural Science Establishment, Rochester, NY). The peat moss and clay were used as received. Samples of dry peat (3 g) or clay (0.4 g) were added to the VOA vials which were then filled with an aqueous solution of MTBE (11.3 mM final concentration) made with distilled water. The pH of the solutions in contact with peat and clay were 3.5 and 4.5 respectively.

RESULTS

The hydrolysis of ether oxygenates is a second order reaction. The rate depends on both pH and ether concentration. A detailed description of the kinetics is presented in O'Reilly et al. (11). In the following results, rates constants are reported as first order for ether concentration at a given pH.

Figure 1 shows the time course of TBA production for incubations conducted with 10, 25, 50 and 100 mM MTBE at pH 2.0. In this experiment TBA was generated from MTBE and the rate of TBA production was constant over time for each concentration of MTBE tested. Furthermore, the rate of TBA production was proportional to the initial MTBE concentration included in each reaction. Based on these observations, a rate constant for the four reactions of 1.0×10^{-3} d^{-1} was determined. Methanol was also detected as a product of MTBE hydrolysis.

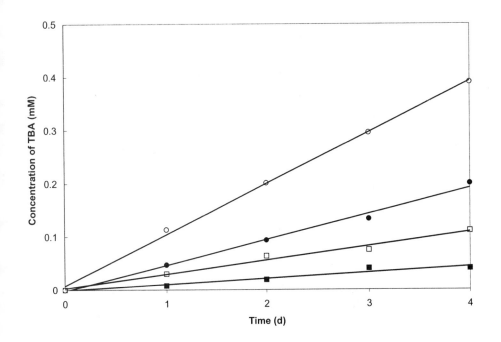

Figure 1- Concentration of TBA generated from (■) 10 mM, (□) 25 mM, (●) 50 mM and (o) 100 mM MTBE in phosphate-buffered reaction media (pH2).

Subsequent experiments examined the effect of pH on the hydrolysis of MTBE and also investigated whether similar hydrolysis reactions occurs with other gasoline oxygenates. The results (Figure 2) demonstrate that MTBE was only hydrolyzed to TBA at substantial rates at pH values less than 3. The rate constant for MTBE hydrolysis at pH 2.0 was again estimated at 1.0×10^{-3} d^{-1} whereas the rate of hydrolysis decreased to 5.9×10^{-4} d^{-1} at pH 2.5, and 1.2×10^{-4} d^{-1} at pH 3. The rates at pH and higher were not sufficient to detect given the experiment method. Slightly higher rates of hydrolysis were also observed for *tertiary* amyl methyl ether (TAME) and ethyl *tertiary* butyl ether (ETBE) in incubations conducted at pH 2.0. (Figure 3). The hydrolysis of TAME resulted in the accumulation of both *tertiary* amyl alcohol (TAA) and methanol as hydrolysis products. The rate of hydrolysis based on *tertiary* amyl alcohol production was calculated to be 2.5×10^{-3} d^{-1}. The hydrolysis of ETBE generated both TBA and ethanol as products. The rate of hydrolysis based on TBA production was calculated to be 1.8×10^{-3} d^{-1}. No hydrolysis of either TAME or ETBE was observed over a 4 day reaction in incubations conducted at pH 7.0.

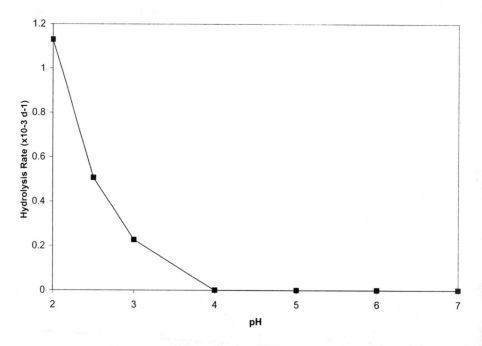

Figure 2- Effect of pH on the rate of MTBE hydrolysis in reactions conducted in phosphate-buffered media with 100 mM MTBE.

Having demonstrated the hydrolysis of MTBE and other ether oxygenates in the liquid phase, we were interested in determining whether similar reactions could be achieved using acidic solid phases. No hydrolysis of MTBE was observed in incubations (up to 21 days) conducted with either peat or montmorillonite clay as the solid phase material. In contrast, rapid hydrolysis of MTBE was observed in the presence of various ion exchange resins (11). The reported rate constant for hydrolysis catalyzed by Amberlite IR-120 resin was 0.67 d^{-1} for a saturated resin slurry at 25 °C.

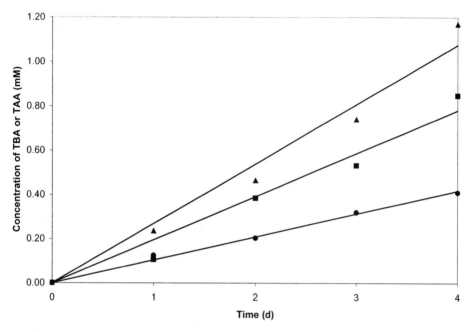

Figure 3- Comparison of the rate of TBA formation from MTBE (●) and ETBE (■) and TAA formation from TAME (▲) at pH 2 and 25 °C. The initial ether concentration was 100 mM.

DISCUSSION

The results presented in this study demonstrate that MTBE and other branched ether oxygenates are susceptible to hydrolysis under conditions that are mild compared to those previously described in the chemical literature. What is less clear from this study is the possible impact of an acid-catalyzed hydrolysis on the environmental fate of MTBE. At this time there is considerable interest in determining whether MTBE undergoes natural attenuation in ground water environments. When evaluating natural attenuation at a site it is important to recognize and understand all of the reactions that may result in mass loss and accurate estimates of transformation rates are required to model plume dynamics over time. The rate of MTBE hydrolysis we have determined at pH 2 (1×10^{-3} d^{-1}) is at the lower end of the range of natural attenuation rates of benzene (12), and similar to rates calculated for MTBE in the subsurface (13). It is well accepted that rates of this magnitude can control plume size. Although we recognize that pH 2 is considerably lower than the typical pH of subsurface environments, our results with ion exchange resins (11) open the possibility that acidic surfaces can catalyze the hydrolysis of MTBE at higher pH values. Given the high cation exchange capacity of many soils, and the high soil surface area to water volume ratio, hydrolysis may be an abiotic mechanism for MTBE attenuation in groundwater. This hypothesis requires further testing. The two natural materials evaluated did not hydrolyze MTBE under the experimental conditions. Although both clay and peat have negatively charged surfaces like acidic ion exchange resin, these are likely saturated with mineral counter ions such as sodium or calcium. The resin required acid activation to saturate the surfaces with protons (11). Geochemical conditions that result in proton rich surfaces would be more amenable to solid catalyzed hydrolysis. Dissolved metals and metal rich minerals can act as Lewis acids. The ability of these to participate in MTBE hydrolysis also needs to be evaluated.

It is important to note that the reactions described in this study lead to the transformation, not complete degradation of ether oxygenates. Understanding the fate of the alcohol products will also be important in assessing the impact of these findings on groundwater protection and possible treatment processes. Methanol should be readily degraded under either aerobic or anaerobic conditions (14,15). In contrast, TBA degrading activity is less well characterized although it has been detected at some (14, 16), but not all (15) sites that have been investigated for biodegradation processes.

Although this discussion has focussed largely on abiotic transformations of MTBE, our results also have potential implications for understanding the biodegradation of ether oxygenates. To date all of the isolated MTBE-degrading microbial cultures are aerobic (17,18) and this is consistent with the need for oxygenase enzymes to cleave the ether bond (19,20). However, there have been some limited indications of anaerobic MTBE degradation (13, 15, 16), although the mechanism of the ether bond cleavage reaction is unknown. Hydrolyases are a recognized class of enzymes (21) that can catalyze hydrolysis reactions under either aerobic or anaerobic conditions. Moreover, many hydrolytic enzymes have an active site that behaves as a low pH environment and have amino acid residues that can serve as the nucleophile required for the proposed hydrolysis reaction. An enzyme-catalyzed hydrolysis has been proposed for cleaving the C-O bond in α-glycosidic reactions (22). White et al. (23) discussed the possibility of a hydrolytic enzymatic mechanism in their review of the biological scission of ether bonds. Although they presented limited examples of ether hydrolyases, they felt that these types of enzyme may not have much importance since, as they state, "ethers are ... highly stable to hydrolysis even in the presence of mild acids". The results presented in this study suggest that ether oxygenates may be sufficiently susceptible to hydrolysis to be subject enzymatic cleavage.

ACKNOWLEDGMENT

A portion of this work was funded by the American Petroleum Institute. The opinions expressed are those of the authors and not necessarily those of the funding agency.

REFERENCES

1. Squillace, P.J.; Pankow, J.F.; Korte, N.E.; Zogorski, J.S. 1998. *Environmental Behavior and Fate of Methyl tert-Butyl Ether (MTBE)*. U.S. Geological Survey., Washington, D.C., 4 pages.

2. Howard, P.H.; Boethling, R.S.; Jarvis, W.F.; Meylan, W.M.; Michalenko, E.M. 1991. *Handbook of Environmental Degradation Rates*. Lewis Publishers, Chelsea, MI.

3. Burrell, R.L., Jr. 1954. "The Cleavage of Ethers." *Chem. Rev.* 54(4): 615-685.

4. Zogorski, J.S., Baehr, A.L., Bauman, B.M., Conrad, D.L., Drew, R.T., Korte, N.E., Lapham, W.W., Morduchowitz, A., Pankow, J.F., and Washington, E.R., 1997. "Significant findings and water-quality recommendations of the Interagency Oxygenated Fuel Assessment." *in* Kostecki, P.T., Calabrese, E.J., and Bonazountas, Marc, (Eds.), *Contaminated Soils*, pp. 661-679. Amherst Scientific Publishers, Amherst, MA.

5. Butlerow, A. 1861, "Mitheilungen aus dem Laboratorium zu Kasan. 1. Ueber die Aethylmilchsäure." *Ann.* 118: 325-330.

6. Silva, R. 1875, "Akademie, Sitzung vom 16 August." *Ber.* 8: 1352-1353.

7. Lippert, W. 1893. "Ueber die Zersetzung der Aether durch Wasserstoffsäuren." *Ann.* 276: 148-199.

8. Baeyer, A. 1893. "Ortsbestimmungen in der Terpenreihe." *Ber.* 26(2), 2558-2565.

9. Norris, J.F., Rigby, G.W. 1932. "The Reactivity of Atoms and Groups in Organic Compounds. XII. The Preparation and Properties of Mixed Aliphatic Ethers with Special Reference to those Containing the Tert-Butyl Radical." *J. Am. Chem. Soc.* 54(5): 2088-2100.

10. Church, C.D., Pankow, J.F., Tratnek, P.G. 1999. "Hydrolysis of tert-Butyl Formate: Kinetics, Products, and Implications of for the Environmental Impact of Methyl tert-Butyl Ether." *Env. Tox. Chem.* 18(12): 2789-2796.

11. O'Reilly, K.T.; Moir M.E.; Taylor, C.; Hyman, M. "Hydrolysis of *tert*-Butyl Methyl Ether (MTBE) in Dilute Aqueous Solutions." *Environ. Sci. Technol.* Submitted.

12. Rifai, H.S.; Borden, R.C.; Wilson, J.T,; Ward, C.H. 1995. In *Intrinsic Bioremediation*; Hinchee, R.E.; Wilson, J.T.; Downey, D.C. (Eds.) pp.1-30. Battelle Press, Columbus, OH.

13. Wilson, J.T.; Cho, J.S., Wilson, B.H.; Vardy, J.A., 2000. *Natural Attenuation of MTBE in the Subsurface under Methanogenic Conditions.* U.S. Environmental Protection Agency, Washington D.C.

14. Hickman, G.T.; Novak, J.T. 1989. "Relationship between Subsurgace Biodegradation Rates and Microbial Density." *Environ. Sci. Technol.* 23(5): 525-532.

15. Mormile, M.R.; Liu, S.; Sulflita, J.M. 1994. "Anaerobic Biodegradation of Gasoline Oxygenates: Extrapolation of Information to Multiple Sites and Redox Conditions". *Environ. Sci. Technol.* 28(9): 1727-1732.

16. Yeh, C.K.; Novak, J.T. 1994. "Anaerobic Biodegradation of Gasoline Oxygenates in Soils." *Water Environ. Research.* 66(5): 744-752.

17. Salanitro, J.P.; Diaz, L.A.; Williams M.P.; Wisniewski, H.L. 1994. "Isolation of a Bacterial Culture that Degrades Methyl *tert*-Butyl Ether." *Appl. Environ. Microbiol.* 60(7): 2593-2596.

18. Hanson, J.R.; Ackerman, C.E.; Scow, K.M. 1999. "Biodegradation of Methyl tert-Butyl Ether by a Bacterial Pure Culture." *Appl. Environ. Microbiol.* 65(11): 4788-4792.

19. Hardison, L.K.; Curry, S.S.; Ciuffett, L.M., Hyman, M.R. 1997. "Metabolism of Diethyl Ether and Cometabolism of Methyl *tert*-Butyl Ether by Filamentous Fungus." *Appl. Environ. Microbiol.* 63(8): 3059-3067.

20. Steffan, R.J.; McClay, K.; Vainverg, S.; Condee, C.W.; Zhang, D. 1997. "Biodegradation of the Gasoline Oxygenates Methyl *tert*-Butyl Ether, Ethyl *tert*-Butyl Ether, and *tert*-Amyl Methyl Ether by Propane Oxidizing Bacteria." *Appl. Environ. Microbiol.* 63(11): 4216-4222.

21. Walsh, C., 1979. *Enzymatic Reaction Mechanisms*. W.H. Freeman and Company, San Francisco, CA.

22. Wallenfels, K.; Diekmann, H. 1967. "Biological formations and reactions." In Patai, S., (Ed.) *The Chemistry of the Ether Linkage*, pp. 207-242. Interscience, London.

23. White, G.F.; Russell, N.J.; Tidswell, E.C. 1996. "Bacterial Scission of Ether Bonds." *Microbiological Reviews.* 60(1): 216-232.

FLUIDIZED BED BIOREACTOR FOR MTBE AND TBA IN WATER

Joseph E. O'Connell – Environmental Resolutions, Inc.
Dallas Weaver – Scientific Hatcheries, Inc.

ABSTRACT: Groundwater, which is contaminated with Methyl Tertiary-Butyl Ether (MTBE) and/or Tertiary Butyl Alcohol (TBA), one of the byproducts of MTBE decomposition, has serious taste and odor problems. Because of their high water solubilities, low Henry's law constants and poor adsorption coefficients onto carbon and soils, MTBE and TBA are difficult to remove from water using conventional methods, such as stripping or adsorption on activated carbon. The biological decomposition of MTBE has been demonstrated, however, essentially all of the species and strains of bacteria capable of oxidizing pure MTBE are considered slow growing. This slow growth creates field startup time problems and competitive exclusion problems and instabilities relating biomass composition in many classes of biological reactors. To solve these problems, Environmental Resolutions, Inc, (ERI) and Scientific Hatcheries have developed a fine media fluidized bed biological reactor specifically designed to stabilize MTBE oxidizing bacteria under mixed culture conditions. The experimental reactors and commercial scale skid-mounted systems installed in the field have been shown to remove both MTBE and TBA down to non-detectable levels (< 1 ug/L and < 5 ug/L respectively) with inlet concentrations in the 1700+ ug/L range (single pass with approximately 15 minute contact time). The fundamental design is fully scaleable to arbitrary sizes and inlet concentrations while being relatively stable (relative to other designs) in the presence of other, more easily degraded, organic pollutants such as BTEX. Active media production reactors have been built and transportation technology developed which make possible rapid startup of the commercial-scale field reactors.

INTRODUCTION

Methyl tertiary-butyl ether (MTBE) has been used as a gasoline additive for at least the past 10 years. In 1994 the EPA required use of an oxygenate in gasoline used in the State of California to reduce the unburned hydrocarbons in auto exhaust, an important contributor to smog. The refiners selected MTBE as the oxygenate of choice and included about 11 – 15% MTBE in current fuel formulations. Shortly after its introduction, MTBE and some of its decomposition products, notably tertiary-butyl alcohol (TBA), began to show up in groundwater in several locations throughout the state. The high water solubility combined with the low absorption by soils resulted in a situation where the MTBE and/or TBA in leaked gasoline entered the ground water and migrated long distances from the source of a gasoline spill.

The presence of low concentrations of MTBE in water (20 to 40 ug/L) produces an unpleasant taste. Concerns about drinking water supplies in California and in other states have prompted regulatory agencies to require the clean up of MTBE and TBA impacted soil and groundwater in sensitive locations throughout the State.

Efforts to remove MTBE and TBA from groundwater and air have focused on physical and chemical processes such stripping, sparging, carbon adsorption, and incineration. These processes have been effective in removing other volatile organic compounds found in gasoline, such as benzene, toluene ethylbenzene and xylene (BTEX). However, the physical properties of MTBE and TBA, such as high water solubility, high affinity for water relative to air (low Henry's Law Constant), and high affinity for water relative to clayey soil (low oil/water partition coefficient, make MTBE and TBA more difficult (and expensive) to remove from groundwater than BTEX materials.

Although MTBE and TBA are known to be biodegradable [1-10] the growth rates of the bacteria involved in the oxidation of MTBE cause them all to be classified as slow growing strains. Some species of bacteria also co-metabolize MTBE along with some other hydrocarbon [11]. Reliable and effective biodegradation of MTBE in the vapor phase has been demonstrated using a vapor phase bio-filter in laboratory and field situations [6].

Scientific Hatcheries has been designing and utilizing fine media fluidized bed biofilters (FBB's) for the oxidation of ug/L levels of ammonia and other organic materials for specialized applications. Over the last 28 years, these biofilter designs have been utilized in hundreds of highly reliable systems capable of removing ammonia from waste waters to non-detectable levels. Some of these systems have 20 years of continuous operation, without shutting down. The design approach for this class of biofilters is oriented towards handling very low substrate concentrations and being stable in the presence of variable amounts of easily biodegradable materials, which support rapid bacterial growth and create competitive exclusion and related problems with nitrifying bacteria. Since the biological aspects of the MTBE removal problem are very similar to those handled by the existing technology, the goal became the modification of the existing reactors for slower growing bacteria and simultaneously adapting bacteria strains, which can oxidize MTBE in the specific biofilter design.

The Experimental Reactor:

A 40 liter/min fluidized bed biofilter was constructed and set up in a recycle mode of operation with a MTBE feed system. The design was modified from the usual ammonia oxidation system designs to account for the slower growth rate of MTBE oxidizing organisms relative to nitrifying organisms.

A culture of MTBE consuming microbes originated from a vapor phase bio-filter operating at the Joint Water Pollution Control Plant in Carson, California. The MTBE-eating microbes were isolated, named PM1, and characterized by researchers at UC Davis. Biodegradation of MTBE in water has been demonstrated for both pure and mixed cultures [8, 12]. Studies at the Oregon Graduate Institute demonstrated that the PM1 culture degraded not only MTBE but also TBA, tert-amyl alcohol (TAA), ethyl tert-butyl ether (ETBE), di-isopropyl ether (DIPE), and tert-amyl methyl ether (TAME)[7]. Starter cultures were also obtained from Dr. Deshusses at UC at Riverside along with soil samples from local gas station cleanup efforts, which had been contaminated with MTBE for long time periods. Therefore, it is expected that the present culture is a mixed culture.

The reactor was setup and seeded in the spring of 1999. Based upon calculations and the estimated growth rate of the PM1 strain, it was estimated that there would be no significant removal for about 150 days. By January of 2000 significant oxygen consumption was seen in the reactor and the exit water no longer tasted of MTBE. Chemical analysis of the MTBE were conducted indicating that the discharge concentrations were below the detection limit of <5 ug/L while the inlet concentrations were in the 1000+ ug/L range.

Growth reactor and prototype commercial reactor:
Reactor #2, the growth reactor, was 1.5 m in diameter and 2.4 m high with a working height of 2 meters with a gravity overflow discharge. The bio-mass was suspended on about 3 metric tons of sand. The recirculation rate was a constant 180 l/min, which gave an empty bed space time of 21 minutes. If allowance is made for media volume, the residence time was about 14 minutes. Effluent from the reactor overflowed through a 60 cm square packed tower for reoxygenation and then into an accumulation tank.

The prototype commercial scale reactor design was similar to Reactor #2 above. The prototype units were skid mounted and had recirculating filters and automatic controls to allow unattended operation for days at a time.

Principles
Oxygen is required for bio-oxidation of organic materials to CO_2 and water. The requirements vary, but for fuel components, about 3 pounds of O_2 are required for every pound of organic material consumed.

In an activated sludge process, a trickling bed filter, and in some fluidized bed reactors, oxygen is added as the reaction proceeds by introducing air into the system by mixing, aspiration, or direct injection. Carbon uptake is limited in part by the mass transfer of oxygen from the air bubbles to the aqueous phase. In fine media fluidized bed biofilters, air bubbles must be eliminated, as they will cause flotation of the media out of the bed. Therefore, bio-oxidation is limited to the amount of dissolved oxygen present in the water entering the bed. Oxygen is introduced in the recycle stream, which is operated at a much higher flow rate than the influent stream, providing a greater capacity for oxygen mass transfer to the reactor.

The equilibrium concentration of dissolved oxygen in water is dependent on the temperature and the partial pressure of oxygen in the gas contacting the water. At sea level, using fresh air (21% O_2) the equilibrium dissolved oxygen is about 9 ppm (by wt) at 72 deg F.

When working with biological reactors, which do not have any nutrient, or oxygen inputs beyond what is contained in the input water supply (e.g. all packed or fluidized bed bio-reactors that do not have gas addition systems) a simple measure of performance is the oxygen consumption. This is the decrease in dissolved oxygen concentration as the fluid passes through the reactor. When working with low substrate concentrations or nutrient sources that have low biomass yield coefficients (don't produce a lot of growth per unit material oxidized), there is a direct relationship between the oxygen consumption and nutrient removal.

Another factor relating to the oxygen levels is the minimum oxygen level required by the organism and the reaction rate as a function of oxygen concentration. With thick biofilm reactors, the reaction rate decreases as the oxygen level decreases as a consequence of the oxygen gradients within the biofilm. With thin biofilm reactors, such as the fine media FBB's used in this work, the reaction rate as a function of discharge oxygen levels, is constant until the minimum oxygen level is reached.

EXPERIMENTAL PROGRAM

Operating Procedure

The growth reactor was charged with sand and put on recirculation in July of 1999. Bio-mass was obtained from UC Davis, UC Riverside and from the pilot reactor operating at Healdsburg, CA. The total bio-mass introduced was estimated to be about 1 gram. A reactor of this size normally requires about 1-2 kg of active bio-mass (an increase of 1000 fold). These bacteria are noted to be slow growers doubling in mass about every 15 days. About ten doubling periods, 150 days, would be required for 1 gram to grow to 1.024 kg (1 x 2^{10} = 1024). Blending grade MTBE (98%) was obtained from the Chevron and Mobil refineries and used to prepare a concentrated feed solution of 4 grams per liter. Feed rates of the concentrated solution were adjusted to keep sufficient MTBE in the recirculating stream to promote growth of the bio-mass. The reactor was sampled periodically to keep tabs on the reaction using procedures outlined below in the section on Sampling and Analytical Procedures.

In February of 2000 a portion of the media from the growth reactor was transferred to the experimental reactor where MTBE feed concentrations and dissolved oxygen concentrations were varied. In April, 500 g of reagent grade TBA was used to prepare a feed solution with both MTBE and TBA. The feed solution, containing 3 grams/liter (g/L) of MTBE and 1 g/L of TBA, was fed into a recirculating stream (38 liters per minute (Lpm)) to give a feed to the reactor of 1300 ug/L MTBE and 400 ug/L TBA. The dissolved oxygen concentration in the reactor feed was about 8 ppm. The solution was introduced to the reactor on April 4th. Samples were taken from inlet and outlet of the reactor and checked for MTBE and TBA on April 5th, 6th and 7th. In January 2001, the TBA augmentation was repeated. This time more TBA than MTBE was fed.

The prototype commercial scale reactor was installed in northern California. After initial system physical startup, the system was put on recirculation and charged with bio-mass July 19, 2000. The "start-up" charge for this reactor was about 1/3 of the biomass existing in the growth reactor. The system was allowed to stabilize for two weeks and the first sample was taken for analysis on August 9, 2000. Samples were taken once every one to two weeks after that time.

Sampling and Analytical Procedures

Each sample was collected in three 40-ml vials with screw caps. Care was exercised to be sure the vials were completely full and there were no air bubbles. The samples were placed in an ice chest and taken to a laboratory certified by the State of California to perform the required analyses. Labs used were Cal Science Laboratories in Garden Grove, CA, Sequoia Laboratories in Redwood City, CA, and Kiff Laboratories in Davis, CA.

MTBE concentrations were determined using a gas chromatograph with a flame ionization detector following EPA Method 8020. The detection limit for this method is 5 ug/L. This method is not used for TBA.

TBA (and also some MTBE) concentrations were determined using EPA Method 8260B (temperature programmed purge and trap and gas chromatograph with a mass spectroscopy detector. Some laboratories (Kiff) claim detection limits of 1 ug/L for MTBE and 5 ug/L for TBA using this method.

RESULTS

The analytical results are summarized in Table 1. Samples taken in July and September suggested more time was required for growth of the microbial mass for effective MTBE removal. Samples taken in February (180+ days after inoculation) indicate sufficient bio-mass is present to reduce MTBE concentration to below 5 ug/L in one pass through the reactor.

TABLE 1
Pilot Plant Analytical Results

Date Sampled	Reactor	MTBE in ug/L	MTBE out ug/L	TBA in ug/L	TBA out ug/L
7/15/99	Growth	52,000	56,000		
7/15/99 (dupl.)	Growth	60,000	62,000		
9/9/99	60 cm	9,230	8,780		
2/10/00	60 cm	680	ND<5		
2/10/00 (dupl.)	60 cm	480	ND<5		
2/10/00	60 cm	190	ND<5		
2/10/00 (dupl.)	Growth	200	ND<5		
2/14/00	60 cm	780	ND<5		
2/14/00 (dupl.)	60 cm	730	ND<5		
3/7/00	60 cm	5,500	3,700		
3/14/00	60 cm	42,000	36,000		
3/23/00	60 cm	1,700	ND<1	None added	ND<5
4/5/00	60 cm	1,300	ND<1	350	ND<5
4/6/00	60 cm	1,300	ND<1	390	ND<5
4/7/00	60 cm	1,100	ND<1	312	ND<5
1/22/01	60 cm	684	ND<1	1,300	ND<5
1/29/01	60 cm	12	ND<0.5	800	ND<5
1/22/01	60 cm	4	ND<0.5	610	ND<5

Blank spaces indicate no analysis was performed

A fluid bed bio-reactor is limited to about 2000 ug/L of MTBE consumption per pass by the amount of dissolved oxygen available in the recirculating stream. The samples taken on March 7 from Reactor #1 and on March 14th from Reactor #2 demonstrated that when the feed contained more than 2000 ug/L of MTBE, the effluent increased dramatically, because the system capacity to consume MTBE was exceeded.

Prior to the March 14th sample from the 60 cm reactor the recirculating stream had been boosted with pure oxygen so the incoming oxygen was about 18 ppm. The result was an MTBE consumption of 6000 ug/L in one pass.

The March 23 sample taken for MTBE analysis via method 8020 was re-analyzed for TBA using EPA Method 8260. TBA was not found indicating that no TBA is produced in this reactor when MTBE is being metabolized.

Samples taken April 5th, 6th and 7th 2000 and the sample taken January 22nd, 2001 indicate both MTBE and TBA were present in the influent to the reactor but were below detectable limits in the effluent. This would indicate that TBA, if found in the groundwater influent, can be readily metabolized by the culture.

Prototype Commercial Reactor

Samples taken from the prototype commercial reactor since its start-up in July are summarized in Table 2 and indicate that MTBE (and TBA) concentrations in the effluent that are less than the limits of detection (ND<2.5 ug/L for MTBE and ND<5 ug/L for TBA).

On 9/22/00 the feed rates were increased to exceed the reactors oxygen capacity, and the DO out of the reactor dropped to less than 1.0 ppm and the concentration of MTBE rose in the reactor effluent. An oxygen booster system was added and the concentrations in the reactor outlet again were below the detection limits.

Samples taken between 10/3 and 10/31/00 indicated some MTBE in the effluent. This was determined to be caused by by-passing in the reactor. Bypassing was corrected by proper baffling and samples taken after 11/9/00 have been <5 ug/L MTBE.

Samples taken 11/9/00, 11/21/00, and 12/20/00 were sent to a lab specializing in detection of low concentrations of TBA in the presence of MTBE (Kiff Labs of Davis California). Although TBA was present in the feed to the reactor at concentrations between 120 and 300 ug/L no TBA was detected in the reactor outlet.

TABLE 2
Prototype Commercial Reactor Analytical Results

Date	Flow into System	Well Influent		Fluid Bed Top		ERI-4000 Exit	
		MTBE	TBA	MTBE	TBA	MTBE	TBA
	L/min	ug/L	ug/L	ug/L	ug/L	ug/L	ug/L
8/9/00	38	6,200		4.74**		ND<2.5	
8/11/00	38	7,240	ND<284	ND<0.5	ND<20		
9/19/00	38	7,600				ND<2.5	
9/22/00	45	13,700		41		30.6	
9/28/00	45	11,100	ND<980	ND<1	ND<3.9	ND<1	ND<3.9
10/3/00	45	10,600	ND<1570	1	ND<3.9	4.1	ND<3.9
10/12/00	53	8,400		ND<2.5		3.5	
10/26/00	64	7,180		ND<2.5		4.56	
10/26/00	64	9,670	ND<980	0.45	ND<3.92	6.11	ND<3.92
10/31/00	64	8,240		22.7		67.5	
11/9/00	64	12,000	300	ND<0.5	ND<5		
11/21/00	64	2,720		ND<2		ND<2	
11/21/00	64	5,300	120			1.2	ND<5
11/29/00	72	5,120		ND<2		ND<2	
12/20/00	72	3,800	260			0.72	ND<5
1/4/01	72	2,870		ND<2		ND<2	
1/24/01	72	6,090		2.39		ND<2	

Blank spaces indicate no analysis was performed

CONCLUSIONS

MTBE, TBA, and presumably other oxygenates, were metabolized by cultures containing PM1 to concentrations that were below the detection limit of current analytical methods (< 5 ug/L) in one pass through a fine media fluidized bed reactor.

Very concentrated solutions (3 g/L for the pilot plant and 12,000 ug/L for the prototype system) were handled by diluting with recycle until the concentration entering the fluidized bed was below that required to completely consume the dissolved oxygen in the recirculating stream.

Field units can be "jump started" using biomass from a growth reactor and can reach full MTBE consuming potential within a few days.

REFERENCES

1. Steffan, R.J., K. McClay, and C.C. Condee, *Biodegradation of methyl tert-butyl ether (MTBE)*. Abstracts of the General Meeting of the American Society for Microbiology, 1997. **97**(0): p. 517.

2. Steffan, R.J., et al., *Biodegradation of the gasoline oxygenates methyl tert-butyl ether, ethyl tert-butyl ether, and tert-amyl methyl ether by propane-oxidizing bacteria*. Applied and Environmental Microbiology, 1997. **63**(11): p. 4216-4222.

3. Stringfellow, W.T. and E.H. Rychel, *Assay of MTBE biodegradation potential in subsurface soils*. Abstracts of the General Meeting of the American Society for Microbiology, 2000. **100**: p. 552.

4. Vainberg, S., R. Unterman, and R.J. Steffan, *Biodegradation of methyl tert-butyl ether (MTBE) and 1,4-dioxane in a fluidized bed bioreactor (FBR)*. Abstracts of the General Meeting of the American Society for Microbiology, 1999. **99**: p. 555.

5. Yeh, C.K. and J.T. Novak, *Anaerobic biodegradation of gasoline oxygenates in soils*. Water Environment Research, 1994. **66**(5): p. 744-752.

6. Eweis, J.B., J.H. Scarano, B.M. Converse, D.P.Y. Chang, E.D. Schroeder, *"Vapor Phase Biodegradation of MTBE,"*. 1999, Center for Environmental & Water Resources Engineering Department of Civil & Environmental Engineering, University of California, Davis.

7. Church, C.D., P.G. Tratnyek, K.M. Scow, *Pathways for the Degradation of MTBE and other Fuel Oxygenates by Isolate PM1*. 1999, Environmental Science and Engineering, Oregon Graduate Institute, Beaverton, OR.

8. Converse, B.M., E.D. Schroeder, D.P.Y. Chang. *Liquid Phase Bio-degradation of Methyl Tertiary Butyl Ether Using a Granular Activated Carbon Trickling Filter in the Field*. in *Paper #714, Session WR-2h, Presented at the 93rd Annual Meeting and Exhibition, June 18-22*. 2000. Salt Lake City.

9. Hanson, J.R., C.E. Ackerman, and K.M. Scow, *Biodegradation of methyl tert-butyl ether by a bacterial pure culture*. Applied and Environmental Microbiology, 1999. **65**(11): p. 4788-4792.

10. Hanson, J.R., C.E. Ackerman, and K.M. Scow, *Biodegradation of MTBE by bacterial strain PM1*. Abstracts of the General Meeting of the American Society for Microbiology, 1999. **99**: p. 552.

11. Garnier, P.M., et al., *Cometabolic biodegradation of methyl t-butyl ether by Pseudomonas aeruginosa grown on pentane*. Applied Microbiology and Biotechnology, 1999. **51**(4): p. 498-503.

12. Fan, M., *Degradation of MTBE by a Suspended Culture*, in *Department of Civil & Environmental Engineering*. 1999, University of California at Davis: Davis.

ENHANCED BIOREMEDIATION OF MTBE IN GROUNDWATER

M. Talaat Balba, Darlene Coons, Rob Hoag, Cindy Lin, Susan Scrocchi, and Alan Weston (Conestoga-Rovers & Associates)

ABSTRACT: A bench-scale column study was conducted to evaluate the potential application of granular humate material for the enhanced bioremediation of methyl tert-butyl ether (MTBE) in groundwater. A glass column was packed with a mixture of granular activated carbon (GAC) and granular humate material, and supplemented with oxygen release compound (ORC). The activated carbon was used to enhance the adsorption of MTBE. Humate was used to stimulate microbial activities and MTBE degradation by providing a slow release source of labile organic carbon and inorganic nutrients necessary for the cometabolism of MTBE. A low concentration of ORC was added to provide a slow release source of oxygen, which is the necessary electron acceptor for the biodegradation of MTBE.

Groundwater contaminated with MTBE was fed to the column from a Teflon bag by the aid of peristaltic pump in a downflow mode. Groundwater influent and effluent were monitored approximately biweekly for the first two months and monthly for the next four months for MTBE concentration to determine removal efficiency of the column. Influent MTBE concentration in the groundwater ranged between 6 and 15.5 parts per million (ppm). Levels in the effluent remained constantly below detection limit (0.01 ppm) for a period of four months of operation (approximately 35 pore volumes). The column packing material was sampled at the end of the study to assess the biofilm microbial characterization and examine the potential MTBE biodegradation and accumulation. The results showed that more than 60 percent of the MTBE removed from the groundwater was degraded by the microbial activities in the column system.

INTRODUCTION

The remediation of hydrocarbon-impacted groundwater at downstream retail and bulk fuel storage facilities is often driven by the presence of dissolved benzene, toluene, ethylbenzene, and total xylenes (BTEX), and more recently by concentrations of dissolved oxygenate MTBE and its biodegradation product tert-butyl alcohol (TBA). In light of the USEPA's recently announced intention to reduce or eliminate the use of MTBE in domestic fuels over the next few years, identification of MTBE attenuation mechanisms is crucial to understand the long-term behavior and final fate of MTBE in the environment. MTBE has a high aqueous solubility, low taste and odor threshold, is tentatively classified by the USEPA as a possible carcinogen, and has a USEPA drinking water advisory of 20-40 μg/L. Typically, these compounds are treated ex situ using air stripping and/or GAC adsorption technology. However, neither of these technologies is very efficient in the treatment of MTBE and TBA. Accordingly, dissolved MTBE and/or TBA concentrations can become the principal drivers for remedial costs (for air stripper size or GAC technologies). Considering the increased regulatory scrutiny and regulation of MTBE and TBA, current and future remediation costs may be significantly increased in order to address dissolved MTBE and/or TBA concentrations. In addition, both of these remedial technologies result in cross-media transfer and do not ultimately destroy the compounds. Bioremediation provides an alternative cost-effective remedial technology for MTBE.

Site Background. The groundwater used in this study was collected from a light industrial site in southeast Michigan where specialty steel products were manufactured. A 10,000-gallon underground storage tank installed in 1986 was used for unleaded fuel storage. An assessment conducted at the site in November 1997 revealed the presence of BTEX/MTBE in groundwater. The tank was removed in December 1997, during which time it was discovered that the tank had been improperly anchored and the fill port had cracked during frost heaving, resulting in release of fuel into the subsurface environment. The duration and extent of the release are unknown since fuel reconciliation records were not maintained. The subsurface soils consist predominantly of silty sands with interbedded clay. Groundwater is perched above a clay confining layer that is present at 10 to 20 feet below ground surface (BGS). Depth to groundwater is 2 to 6 feet BGS. Ongoing remedial efforts involving air sparging/soil vapor extraction (SVE) resulted in significant decline in BTEX concentrations but MTBE degradation is much slower. The main objective of the treatability study was to explore a cost effective innovative treatment alternative for treatment of the MTBE in the groundwater.

OBJECTIVE

This treatability study was conducted to determine the feasibility of using granular humate materials to enhance bioremediation of groundwater containing MTBE. Humates are low-cost, naturally occurring organic materials that are extracted from the ground and pulverized into a granular form. Humates are known to stimulate biodegradation of organic hydrocarbons in wastewater by acting as a source for releasing carbon and nutrients. If successful, humates could potentially be used as either an in situ or ex situ remediation technique to replace or supplement existing remedial technologies and significantly reduce overall remediation costs.

The treatability study was conducted at Conestoga-Rovers & Associates' (CRA's) Treatability Laboratory in Niagara Falls, New York, using representative groundwater collected from the MTBE-impacted site described above.

MATERIALS AND METHODS

Groundwater Characterization. Groundwater samples were characterized for BTEX and MTBE as well as key pertinent parameters for bioremediation, including pH, total and specific microbial counts, nitrate, ammonia, total nitrogen, phosphate, total suspended solids, and soluble metals using SW-846 methods.

Experimental Setup. A glass column (5cm in diameter x 30cm in height) was packed with granular humate material/activated carbon/ceramic chips (1:1:1 v/v/v), and 1 percent magnesium peroxide by weight. A stopcock was used at the bottom of the column to provide adjustable flow control. The experimental setup is shown in Figure 1.

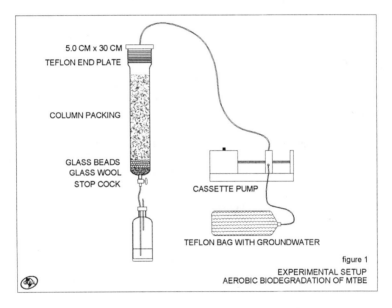

FIGURE 1. Experimental Setup of MTBE Column

The column was first washed with distilled water until no fine particles passed through (i.e., the effluent was clear). The column was then inoculated with the addition of a 100-mL sample of an enriched microbial culture that was established in mineral salt media and MTBE, which was the sole source of carbon and energy. The inoculated column was kept static for 48 hours to allow for microbial attachment. The MTBE-contaminated groundwater was fed into the column in a downflow mode from a Teflon bag using a peristaltic pump. The feed rate was 80-100 mL per day. The bed volume of the column was approximately 420 cm^3. The porosity of the packing material was estimated at 25 percent and the pore volume of the bed was calculated at 105 mL.

The laboratory treatability study was run for approximately 8 months. The column operated five days a week for approximately 6 hours each day.

SAMPLING AND ANALYSIS

Groundwater samples were collected approximately biweekly for the first two months and monthly for the next four months from both the feed and effluent and analyzed for MTBE. The column effluent was also periodically monitored for the key parameters pertinent to biodegradation, including pH, total and specific microbial counts, nitrate, ammonia, total nitrogen, phosphate, total suspended solids, and soluble metals, to ensure that the levels were sufficient for biodegradation. During the last month of the study, the feed and effluent were analyzed weekly for BTEX and TBA concentrations. At the end of the study, samples of the column packing material were collected from the top, middle, and bottom of the column. These samples were examined for total counts of heterotrophic microbial population and the specific MTBE-degrading microbial population. The samples were also analyzed for concentration of residual MTBE concentration to assess MTBE removal by biodegradation and adsorption.

RESULTS AND DISCUSSION

The initial concentration of MTBE in the groundwater was 15.5 mg/L. The concentration in the feed throughout the study fluctuated between 6-14 mg/L. The final concentration in the effluent from the column dropped to 2 mg/L after 1.4 bed volumes of groundwater had passed through the column (see Table 1). The effluent concentration was further reduced to near or below the detection limit (0.01 ppm) after 2.1 bed volumes had passed through. Greater than 99 percent MTBE removal was maintained for 20 bed volumes (see Table 1).

TABLE 1. MTBE Concentrations (in mg/L) in Effluent from Column

Day	Bed Volume	Initial Concentration	Final Concentration	% Reduction
0	0	15.5	-	-
10	1.4	12	2.0	83
15	2.1	6.0	0.092	98
30	3.8	8.9	0.039	> 99
40	5.4	NA	0.030	-
50	6.3	11	0.017	> 99
70	8.1	8.6	0.01	> 99
100	11.1	7.27	ND 0.01	> 99
140	16	6.6	0.004	> 99
180	17.7	14	0.01	> 99
200	20	7.5	0.06	> 99

Notes:
MTBE = Methyl tert-butyl ether.
NA = Not available.
ND = Non-detect at associated value.

The cumulative MTBE concentration in the column was calculated and is shown graphically in Figure 2. Approximately 72 mg of MTBE was removed from the groundwater after 20 pore volumes of groundwater had passed through the column. The residual MTBE concentrations detected in the effluent are also shown in Figure 2.

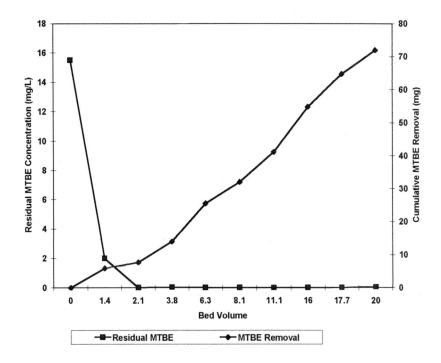

FIGURE 2. Graph of Residual MTBE Concentration Observed in Effluent and Cumulative MTBE Concentration in Column

The column treatment also reduced the concentrations of BTEX compounds from an average concentration of 5 ppm in the feed to non-detect at 0.01 ppm in the effluent. The TBA concentrations in the influent averaged less than 1 ppm and the effluent concentrations were non-detect at 0.01 ppm.

Ammonia nitrogen and orthophosphate phosphorus were monitored and found to be within the ranges appropriate for biodegradation. The humate material used in this study was rich in these nutrients, so the system had adequate supply of these nutrients. pH remained within the neutral range (7-8).

At the end of the study, the packing material was removed and analyzed for MTBE, BTEX, and microbial count. Table 2 summarizes the results of the corresponding levels of adsorbed BTEX and MTBE on different column depths. The microbial counts of the packing material samples collected from the four different parts of the column showed that the total microbial population throughout the column was on the order of 10^4 cells/g. MTBE-, TBA-, and BTEX-specific microbial populations were each approximately 5-10% of the total population.

TABLE 2. Concentrations (in mg/Kg) in Packing Material

Parameter	0-7 cm*	7-12 cm*	12-17 cm*	17-22 cm*
MTBE	210	24	19	2.6
Benzene	1.2	0.2	0.09	ND
Toluene	0.57	0.18	0.10	0.06
m/p-Xylene	0.32	0.09	0.06	ND
o-Xylene	0.13	ND	ND	ND

Notes:
* = Distance (in centimeters) from top of column.
MTBE = Methyl tert-butyl ether.
ND = Non-detect at 0.01 ppm.

The concentrations of all compounds decreased with depth from the top of the column. This is due to the gradual adsorption of these compounds onto the column and the degradation of these compounds by the microbial-specific bacteria.

It was shown above that approximately 72 mg of MTBE was removed from the groundwater onto the column. The total MTBE mass adsorbed on the column packing material (as calculated from the concentrations shown in Table 2) was estimated at 29.6 mg based on the MTBE residual analysis. Therefore, the results indicate that more than 60 percent of the removed MTBE was degraded by biological activity in the column.

The treatability study is continuing to fully assess the biodegradation kinetics of MTBE in the column system. The flowrate will be gradually increased to assess the maximum loading rate of MTBE and determine the breakthrough point.

SELECTED REFERENCES

Bradley, P.M., J.E. Landmeyer, and F.H. Chapelle. 2001. *Environ. Sci. Technol.* 35: 658-662.

Bradley, P.M., J.E. Landmeyer, and F.H. Chapelle. 1999. *Environ. Sci. Technol.* 33: 1877-1879.

Hutchins, S.R., W.C. Downs, J.T. Wilson, G.B. Smith, D.A. Kovacs, D.D. Fine, R.H. Douglass, and D.J. Hendrix. 1991. *Ground Water 29*: 571-580.

Johnson, R., J. Pankow, D. Bender, C. Price, and J. Zogorski. 2000. *Environ. Sci. Technol. 34*: 210-217.

Reuter, J.E., B.C. Allen, R.C. Richards, J.F. Pankow, C.R. Coldman, and J.S. Seigfried. 1998. *Environ. Sci. Technol. 32*: 3666-3672.

Salanitro, J.P., G.E. Spinnler, C.C. Neaville, P.M. Maner, S.M. Stearns, P.C. Johnson, and C. Bruce. 1999. "Demonstration of the Enhanced MTBE Bioremediation (EMB) In Situ Process." B.C. Alleman and A. Leeson (Eds.). Battelle Press, Columbus. Vol. 3, pp. 37-46.

Squillace, P.J., J.F. Pankow, N.E. Korre, and J.S. Zogorski. 1997. *Environmental Toxicology and Chemistry 16:* 1836-1844

Squillace, P.J., J.F. Pankow, N.E. Korre, and J.S. Zogorski. 1996. *Fact Sheet FS-203-96.* U.S.G.S.

Squillace, P.J., D.A. Pope, and C.V. Price. 1995. *Fact Sheet FS-114-95.* U.S.G.S.

United States Environmental Protection Agency. 1996. *Drinking Water Regulations and Health Advisories.* Washington, D.C.

AEROBIC BIOREMEDIATION OF MTBE AND BTEX AT A USCG FACILITY

P. Hicks (ARCADIS, Raleigh, North Carolina)
M.R. Pahr (ARCADIS, Raleigh, North Carolina)
J.P. Messier (USCG, Elizabeth City, North Carolina)
R. Gillespie (REGENESIS, Plano, Texas)

Abstract: A full-scale demonstration *in-situ* bioremediation of dissolved methl-tert-butyl-ether (MTBE) and benzene, toluene, ethylbenzene and xylenes (BTEX) was successfully completed in 2000 at the U.S. Coast Guard (USCG) Support Center in Elizabeth City, North Carolina. The goal of the project was to enhance the natural biodegradation of dissolved petroleum constituents without interfering with facility operations. The project goal was achieved by aquifer oxygenation which enhances natural bioremediation rates. Aquifer oxygenation was accomplished by injecting Oxygen Release Compound (ORC) into both the source and dissolved plume areas. Post-treatment monitoring of the aquifer indicated a substantial decrease in dissolved MTBE and BTEX concentrations. The dissolved MTBE mass was reduced 100% in both the source and plume areas. The dissolved BTEX mass was reduced 99% in the source area, and 53% in the plume area following oxygenation of the aquifer. Current monitoring efforts verify that bioremediation is occurring at a rate that is sufficient to protect down gradient receptors. Site closure has been obtained from the North Carolina Department of Environment and Natural Resources, Division of Environmental Management.

INTRODUCTION

The USCG retained ARCADIS Geraghty & Miller to identify and execute proactive and innovative technologies for corrective action at a former fuel storage area at the Support Center facility in Elizabeth City, North Carolina. Historical activities resulted in the release of petroleum beneath a former underground storage tank (UST). Planned expansion of the runway system at the facility will render the area inaccessible, and the USCG requested a remedial approach that would enhance attenuation of dissolved petroleum constituent mass without interfering with facility operations or compromising the integrity of the new runway/apron.

Objective. The overall goal of the project was to temporarily enhance the attenuation rate of MTBE and BTEX. This would then shorten the overall project life by reducing concentrations either to acceptable levels during long-term monitoring efforts or below the Title 15A North Carolina Administrative Code Subchapter 2L standards. *In-situ* bioremediation of the residual dissolved petroleum was selected as the most efficient technology and remedial approach. Based on historical experience, enhanced oxygenation of the aquifer via ORC injection was anticipated to achieve the project goals, and degrade both MTBE and BTEX to acceptable levels (Boyle et al., 1999; Hicks et al., 1999; Koenigsberg et al., 1999).

Site Description. The USCG facility is located adjacent to the southern bank of the Pasquotank River, and the primary functions provided at the facility include support, training, operation and maintenance associated with USCG-operated aircraft. These functions have required jet-fuel storage, handling and refueling systems. The treated area previously contained a 10,000-gallon fiberglass UST and a series of underground transfer lines used to deliver JP-4 and JP-5 jet fuel to aircraft at the facility. The majority of the petroleum mass that was inadvertently released in the area had been previously removed during corrective actions implemented during closure of the tanks and lines. Depth to groundwater is approximately 2 m.

The treated area was divided into two sections to facilitate the treatment design and monitoring efforts. The source section is defined as the area in the former tank location, and is approximately 18.3 m wide and 36.5 m long. The plume section is defined as the area directly down gradient of the source, and is approximately 36.5 m wide and 62 m long.

MATERIALS AND METHODS

A phased approach to ORC injections was proposed to maximize the operational flexibility of the remediation process, and to minimize interference with facility operations. Direct push technology facilitated the injection process, and was sufficiently flexible to allow for quick evacuation of the treatment area if required due to aircraft maintenance and search-and-rescue operations. Each ORC application was designed to satisfy oxygen requirements for MTBE and BTEX biodegradation and other naturally occurring oxygen sinks in the aquifer.

ORC Design. Biological oxygen demand (BOD) and chemical oxygen demand (COD) analytical results were used to estimate the total oxygen required per injection event. The software provided by REGENESIS for ORC designs was used to specify the amount of ORC slurry for each injection point. The input for the majority of the design software is readily available from standard monitoring results. The BOD and COD analyses were used to bracket the lower and upper limits of the additional demand factor (ADF) in the design software. Each of these values was divided by the BTEX and MTBE concentrations. The BOD ratio was determined to represent the minimum acceptable ADF value, and the COD ratio was selected as the maximum ADF value.

The initial injection design included 18 points in the source area, and 13 points in the plume area. The 18 source area points each received approximately 16 kg of ORC suspended in 48 l of water, and the 13 plume area points each received approximately 10 kg of ORC suspended in 30 l of water. The secondary treatment design included 16 injection points in the plume area; each received approximately 15 kg of ORC suspended in 45 l of water. More ORC was initially injected in the source area because a greater residual and dissolved petroleum constituent mass was anticipated to be present compared to the plume area. Previous experience with this technology indicated that concentrating more ORC in the source area would increase the potential to achieve the project goals.

Remediation Monitoring. A monitoring program was implemented to measure the efficacy of the oxygenation process. The parametric coverage included quantitative analyses of the dissolved petroleum constituents and both field and laboratory analyses of attenuation factors (oxygen, nitrate, sulfate, and methane). Aquifer characteristics, such as depth to water, temperature, pH, specific conductivity, turbidity and Eh, were also measured during monitoring events.

Monitoring procedures were also designed to not encumber facility operations, and included temporary wells installed with direct push technology. Temporary monitor wells were previously installed in the source area and in the plume area during the investigation phase of the project. To complete the monitor well array, four temporary monitor wells were installed approximately every 15.2 m from the source area through the center of the plume area. The temporary wells were used primarily to monitoring the efficacy of the ORC treatment, but uniform monitoring of all wells was not possible during each event. Figures 1 and 2 show the locations of the monitor wells in both the source and plume areas and the baseline concentrations of MTBE and benzene.

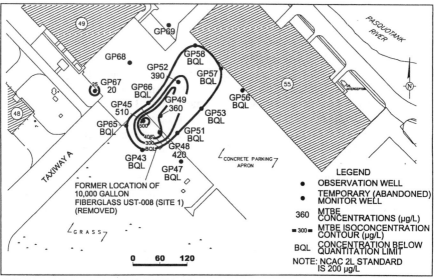

FIGURE 1. PRE-ORC MTBE CONCENTRATIONS (µg/L)

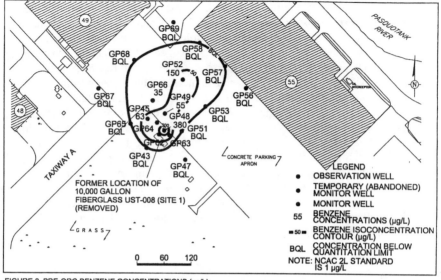

FIGURE 2. PRE-ORC BENZENE CONCENTRATIONS (µg/L)

RESULTS AND DISCUSSION

The data collected during monitoring events following ORC injection indicated that the bulk aquifer appeared to have shifted from uniform anaerobic conditions to a mixed aerobic and anaerobic nature. Background dissolved oxygen (DO) concentrations in the uncontaminated portions of the aquifer ranged between 1.15 mg/L and 6.22 mg/L. Prior to treatment, DO concentrations in the MTBE and BTEX contaminated portions of the aquifer ranged between 0.16 mg/L and 0.64 mg/L.

Following treatment, the DO concentrations in the treated area increased moderately to a maximum of 0.89 mg/L in the source area and to a maximum of 1.72 mg/L in the plume area. The DO levels were generally elevated following ORC injection; however, most of the monitoring points in the treated area did not achieve DO concentrations equal to background conditions.

Similarly, the oxidation-reduction potential (Eh) readings indicated a trend toward more aerobic conditions following ORC injection. The background Eh concentrations in the uncontaminated portions of the aquifer ranged from 110 mV to 143 mV. Prior to treatment, the Eh in the MTBE and BTEX contaminated portions of the aquifer ranged between -47 mV and -428 mV. The Eh in the source area following treatment ranged between –96.4 mV and –120.1mV, and ranged between 8.2 mV to –37.2 mV in the plume area. Nitrate and sulfate concentrations remained below quantitative limits. Temperature, specific conductivity, and pH did not substantially change following treatment. Methane concentrations remained elevated in the source area (3,300 mg/L) and portions of the plume area (<8.6 mg/L to 3,100 mg/L). Although high DO concentrations and Eh conditions were not measured throughout the aquifer, sufficient oxygen was apparently delivered to the aquifer to enhance the rate of biological degradation.

Post treatment monitoring of the aquifer quality parameters indicated effective enhancement of biological activity and substantial decrease in dissolved MTBE and BTEX concentrations. Maximum MTBE concentrations of 510 micrograms per liter (μg/L) in the source area and 390 μg/L in the plume area were recorded prior to treatment. The MTBE concentrations in both source and plume areas decreased to below quantifiable limits approximately 90 days after treatment. The dissolved MTBE mass was reduced 100%, and subsequent monitoring events conducted at 3 months, 6 months, and 7 months after the initial treatment have shown no rebound in MTBE concentrations. Figure 3 shows MTBE concentrations following treatment.

Maximum benzene and BTEX concentrations of 380 μg/L and 745 μg/L, respectively, in the source area and 150 μg/L and 1,100 μg/L, respectively, in the plume area were recorded prior to treatment. The benzene and BTEX concentrations in the source area were reduced to a minimum of 4.3 μg/L and 14.3 μg/L, respectively, approximately 60 days after treatment. The benzene and BTEX concentrations in the plume area were reduced to a minimum of 78 μg/L and 87.5 μg/L, respectively, approximately 90 days after treatment. Some rebound of BTEX concentrations was measured after the initial treatment. Figures 4 and 5 show benzene concentrations in August and December, 2000, following the initial treatment.

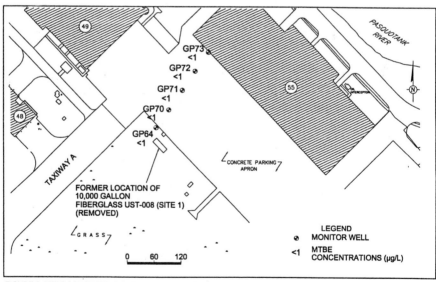

FIGURE 3. MTBE CONCENTRATIONS (μg/L) DECEMBER 2000

112   *Bioremediation of MTBE, Alcohols, and Ethers*

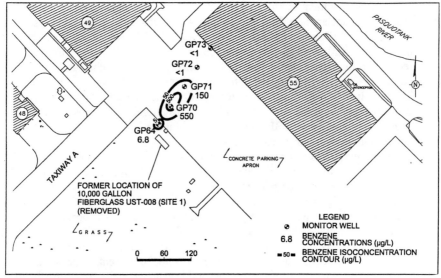

FIGURE 4. BENZENE CONCENTRATIONS (µg/L) AUGUST 2000

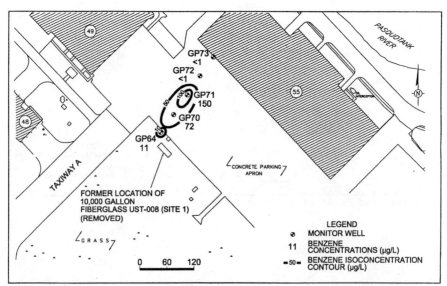

FIGURE 5. BENZENE CONCENTRATIONS (µg/L) DECEMBER 2000

In Situ MTBE Biodegradation

The total dissolved BTEX mass was estimated for each monitoring event by calculating the volume of groundwater in the source and plume areas and then calculating the mass of BTEX contained in each aquifer volume using the concentration data from monitoring points located in each aquifer volume. Figure 6 shows the reduction in benzene mass over time following treatment in the source and plume areas. The BTEX mass was reduced 99% in the source area and 53% in the plume area.

The results of the monitoring events indicate that the ORC design may have produced conservative oxygen loading in the source area and insufficient oxygen loading in the plume area. Residual adsorbed petroleum mass associated with petroleum product release contributes to long-term dissolution and usually requires follow-on ORC injections. In this case, the initial ORC injection was effective in the source area, but the plume area required a second application. It is also possible that some undetected residual adsorbed petroleum mass had migrated below the runway apron and had not been removed during tank closure. This undetected residual mass may have contributed to the slight apparent rebound in dissolved BTEX concentrations in the plume area following the initial ORC treatment.

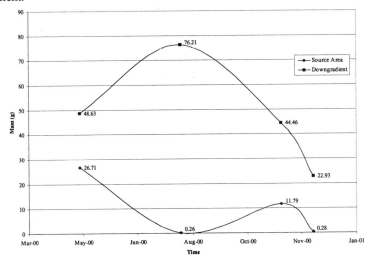

FIGURE 6. BENZENE MASS vs. TIME

Current monitoring efforts verify that remediation is occurring at a rate that is sufficient to protect down gradient receptors. The relative slow groundwater velocity, distance to the Pasquotank River, and absence of dissolved MTBE and reduced BTEX down gradient from the plume area contribute to the success of this project. Site closure has been obtained from the North Carolina Department of Environment and Natural Resources, Division of Environmental Management. Monitoring is anticipated to be suspended in the second quarter of 2001.

REFERENCES

Boyle, et al., 1999. Enhanced *In-Situ* Bioremediation of Groundwater at Madil Air Force Base, FL. in Accelerated Bioremediation Using Slow Release Compounds, Selected Battelle Conference Papers, 1993-1999. Regenesis Bioremediation Products, San Clemente, CA., pp. 163-168.

Hicks, et al., 1999. Combining HVME and Enhanced Bioremediation for Gasoline Contamination in Groundwater. in Accelerated Bioremediation Using Slow Release Compounds, Selected Battelle Conference Papers, 1993-1999. Regenesis Bioremediation Products, San Clemente, CA., pp. 151-156.

Koenigsberg, et al., 1999. Peroxygen Mediated Bioremediation of MTBE. in Accelerated Bioremediation Using Slow Release Compounds, Selected Battelle Conference Papers, 1993-1999. Regenesis Bioremediation Products, San Clemente, CA., pp. 139-144.

FIELD DATA REGARDING AIR-BASED REMEDIATION OF MTBE

Bradford G. Billings (Billings & Associates, Inc., Albuquerque, N.M./USA)
James E. Griswold (Construction Analysis & Management, Inc., Albuquerque, N.M./USA)

Abstract: Several fuel hydrocarbon release sites located in the southwestern United States were selected for analysis of dissolved-phase MTBE remediation. Each site employed biosparging with vacuum extraction as the remedial tool in varied lithologies and depths to water. Specific detail was given toward evaluating removal rates by volatilization and bioremediation. Within three months to two years of operation, direct volatilization rates from 0.002 to 2.18 kg/day for MTBE were inferred. Average dissolved-phase reductions in MTBE ranging from 38 percent to 100 percent are seen. When data were available, a comparison of benzene remediation rates is included. "Age of release" and soil characteristics at MTBE contaminated sites may be highly integral in the rate and dominant process (volatilization versus bioremediation). Which will prove most productive at a given site using sparging technology is undefined at this time.

INTRODUCTION

Methyl tertiary butyl ether (MTBE) is a fuel additive used extensively through the late 1980's and the early 1990's as an oxygenating compound in gasoline. MTBE has a high mobility and solubility relative to other fuel components. Its' vapor pressure is fairly high compared to other gasoline constituents, and its' Henry's law constant is relatively low. One would expect low concentrations of MTBE adsorbed to soil in the vadose zone. This might explain the "expected" poor extraction rates from water using vacuum extraction systems alone. Until recently MTBE was generally considered to have a low potential for biodegradation. Some lag time for MTBE-degrading bacterial growth may exist before supplied oxygen can be more aggressively utilized for effective bioremediation.

Site data used herein were collected from private-sector commercial applications of biosparging technology. In most instances, assessment data from such sites do not lend themselves to an accurate evaluation of initial contaminant mass. Data quantity and type is typically limited by fiscal considerations. Information was collected from six remediation sites with lithologies ranging from fairly homogenous sands, layered sands of varied grain size, gravels and silt, to clay. Depths to water range from 2.0 m to 6.0 m. Relevant data were collected over periods from as brief as the first two months of operation to as long as two years.

OBJECTIVE

An attempt was made to separately assess the volatilization and bioremediation rates of MTBE at several sites employing air-sparging/vacuum extraction, with ambient air being the only injected medium. Rate calculations are based on concentration data collected from the vented air streams of the extraction subsystems along with observed reductions in dissolved-phase contamination. Calculations for direct removal are based on laboratory analysis of bag samples combined with field measured flows. Biologic rates are calculated using oxygen and/or carbon dioxide concentrations in the extraction air streams determined by field instrumentation.

METHOD

System Parameters. In all cases, the remedial tool was biosparging using ambient surface air, coupled with vadose zone vacuum extraction. Spacing between air injection points varied from about 3.0 m to 6.0 m depending on site specific parameters and logistical requirements. In general, the tops of the injection screens were placed from 2.5 to 3.0 m below the encountered water table. Some variation on this depth occurred depending upon the lithology encountered at each site. The extraction screen interval was typically 1.5 m in length. For the vapor extraction system, the depth of placement of the extraction screens was site dependent and horizontal spacing of the screens was coincident with that of the injection points.

Air injection pumps were all rotary-lobe positive displacement units. Extraction pumps were regenerative blowers. All were sized for specific site needs, including cost. Injection and extraction flows varied by site but ranged from 100 and 200 L/sec.

Site Descriptions. The Tuba site is located in northeastern Arizona within the Navajo Reservation and is primarily composed of medium to fine sands to a subsurface depth of approximately 6.0 m. Below this depth, a confining sandstone is encountered. Depth to water is approximately 2.8 m. Due to toxicological considerations raised by the USEPA, and cultural concerns of the Navajo and nearby Hopi populations, extensive chemical analyses of the vented effluent stream were made on a daily basis for the first several months of biosparging operation. The authors felt the increased system monitoring could facilitate an opportunity to better understand the remedial process as well.

The Canon site is located in south-central Colorado and exhibits a multi-layered lithology with silts and clays intermixed in the upper 2.4 to 3.0 m. Medium to coarse-grained sands are then found to a depth of approximately 6.0 m. From 6.0 to 7.5 m a coarse sand and cobble bed is encountered. Depth to water is approximately 6.2 m.

The Beene site is situated in metropolitan Dallas, Texas and is composed entirely of clay with minor silts throughout the soil column. Depth to water averages about 3.5 m.

The Bloom site, located in northwestern New Mexico, is primarily composed of fine to medium sands with occasional silt stringers of irregular thickness. Depth to water is about 2.2 to 2.8 m.

The Herrera site is in metropolitan Albuquerque, New Mexico. The subsurface is composed of primarily alluvial sands of medium to fine grain with intermittent thin silt layers. Depth to water ranges from 3 to 3.5 m across the contaminated area.

The Vagabond site is located in central New Mexico and is dominantly composed of medium sands with occasional silt. Depth to water is approximately 2.4 m below surface.

RESULTS

Dissolved-phase MTBE Analyses. Initial MTBE concentrations in ground water assessed in numerous wells from each site ranged from several hundred µg/L to more than 100,000 µg/L. Data from monitoring wells on the lateral edges of the dissolved-phase plumes were not included in the following table and remained at non-detectable levels throughout. The number of wells used to determine each of the following site averages ranged from 9 to more than 20.

TABLE 1. Site specific dissolved-phase MTBE reduction.

Site	# of Wells	# of Samples	Average Initial Concentration (µg/L)	Operational Time (months)	Dissolved-Phase MTBE Reduction (%)
Tuba	9	27	7,938	3	67
Canon	26	108	13,633	12	79
Beene	12	169	44,386	12	54
Bloom	13	221	4,151	24	88
Herrera	17	289	17,323	24	100
Vagabond	18	108	5,135	6	38

Some wells at these sites did not show any reduction in the concentration of dissolved-phase MTBE and a few wells at some sites revealed modest increases in MTBE levels through the course of remedial operations. Water quality information from these wells was used when estimating the overall percent reductions shown in Table 1.

Analyses of Vapor Extraction Off-Gas. MTBE concentrations in the extraction air streams were measured at only two of the sites: Tuba and Canon. Concentrations ranged from approximately 31 µg/L at Canon to 2,000 µg/L at Tuba. The resultant direct (volatilized) MTBE removal rate at the Canon site was calculated to be approximately 0.0002 kg/day and 2.2 kg/day at Tuba. For comparison purposes, the volatilized benzene removal rate at Canon was 0.005 kg/day and 1.6 kg/day at Tuba.

Using the observed concentration of carbon dioxide in the vent stream along with the measured air flow, the mass of hydrocarbon degradation was estimated in benzene equivalents. These estimates of bioremediation rates are in all instances far higher than the amounts of contaminant directly sparged and/or vented.

Table 2. Estimated Removal Rates

Site	Affected Volume (m³)	Benzene Venting Rate (kg/day)	MTBE Venting Rate (kg/day)	Estimated Biodegradation Rate (kg/day)
Tuba	11,000	1.62	2.18	19
Canon	17,000	0.005	0.002	40
Beene	14,000	0.014	--	62
Bloom	12,000	0.003	--	3
Herrera	13,000	0.009	--	26
Vagabond	16,000	0.004	--	71

Comparison of Venting and Bioremedial Processes. An expected initial residual mass of about 20 kg of MTBE was calculated at the Tuba site based on data from 28 soil borings advanced within the area of the remedial system prior to the author's involvement along with the initial dissolved-phase MTBE information. The directly vented mass of MTBE removed in the first 79 days of system operation approached 200 kg. The authors are unable to distinguish the contributions to the observed cleanup by venting or bioremedial processes separately. The soil data was inadequate to estimate the residual mass of MTBE residing at or near the water table. Consequently, measured changes in contaminant concentrations did not provide sufficient information to determine vented vs. bioremediated contributions. However, the venting process is likely dominating during this early phase of remediation.

The Canon site is an older site than the Tuba site. Available information leads the authors to infer that the release at the Canon site might have begun as early as the 1960s, while for the Tuba site it appears that the first release might have occurred in 1987 or 1988. Nonetheless, both sites have shown significant MTBE contaminant reduction in the groundwater. The MTBE rate of removal at the Canon site is very much less than that observed initially at the Tuba site. The carbon dioxide in the vent stream was much higher at the Canon site (average concentration of 1.01%) than that seen at the Tuba site (0.62%). Therefore, it is plausible that bioremediation contributed more to the overall reduction in dissolved-phase MTBE concentrations at the Canon site than at the Tuba site.

DISCUSSION

Ground water reductions resulting from air sparging/vacuum extraction remedial systems applied to numerous sites with differing lithologies showed a range of 38 to 100 % MTBE reduction. Biosparging significantly reduced MTBE concentrations. Elapsed time since release, the lithologys' influence on the availability of indigenous microorganisms, as well as the duration of system operation all likely played distinct roles in the observed reductions.

Some sites maintained significant rates of remediation for MTBE in the ground water, yet some of these same sites did not reveal high concentrations of MTBE or benzene in the collected air streams compared to other sites. There are indications that despite decreasing MTBE concentrations in the air stream as the remedial systems remained in operation, MTBE concentrations in the ground water continued to decline. There are also indications that carbon dioxide levels remained at significant levels indicating sustained biological activity throughout system operation.

ACKNOWLEDMENTS

The authors wish to thank the following for their continued assistance in these efforts; Kena Fox-Dobbs of Environmental Resource Management for her work in compiling and interpreting the vent data at the Tuba site, Terry Griffin of BioTech Remediation for coordination of groundwater monitoring at both the Tuba and Canon projects, Hall Environmental Analysis Labs for their work on vented air samples, and Mack Brice of the Diamond Shamrock Corporation.

FULL-SCALE REMEDIATION OF AN MTBE-CONTAMINATED SITE

James J. Kang, Ph.D., P.E. (URS Corporation, Santa Ana, California)
Patton B. Harrison, P.E. (American Airlines, Fort Worth, Texas)
Michael F. Pisarik, P.E. (New Fields, Dallas, Texas)

ABSTRACT: The groundwater beneath a former American Airlines UST fuel farm site at JFK International Airport, Jamaica, New York is impacted with high levels of MTBE and moderate levels of BTEX and SVOCs. We found that MTBE is very amenable to advanced oxidation using hydrogen peroxide and ozone. The rate of MTBE degradation seems to follow a pseudo first order reaction. We also found that the dissolution of ozone in water is vital to the destruction of MTBE. Byproducts identified using GC/MS during the reaction include tertiary butyl alcohol (TBA), acetone, and methyl acetate. These byproducts can be either destroyed at the later stage of the reaction or by biological treatment. This paper presents the results of the bench- and pilot-scale treatability studies, as well as the field implementation at the JFK International Airport site.

INTRODUCTION

Methyl tertiary butyl ether (MTBE) is known as one of the fuel oxygenates that has been used in parts of the United States since early 70s (Barker et al., 1990). When used in reformulated gasolines, MTBE reduces ambient exposures to carbon monoxide, ozone (O_3) precursors, and other toxic chemicals that include benzene, a known human carcinogen. Since the Clean Air Act Amendments enacted in 1992, gasoline oxygenate additives such as MTBE, ethyl tertiary butyl ether (ETBE), and tertiary butyl alcohol (TBA) have been added to gasoline products. Of the three gasoline oxygenates, MTBE is the most favored. Reformulated gasoline contains up to approximately 15% MTBE by volume.

Due to its high water solubility and low Henry's law constant, MTBE is difficult to treat when it gets into the groundwater. In addition, MTBE is not readily biodegradable (Mormile et al., 1994; Suflita and Mormile, 1993). Reports indicate that this may be due to the cleavage of the ether bond, which is the rate-limiting step in the biodegradation of MTBE. However, recent studies suggest that MTBE is biodegradable under aerobic or anaerobic conditions using a mixed or pure MTBE degrading culture (Mo et al., 1997; Salanitro et al., 1994; Yeh et al., 1994).

The groundwater beneath a former American Airlines UST fuel farm at John F. Kennedy (JFK) International Airport (the JFK site), Jamaica, New York is impacted with high levels of MTBE and moderate levels of benzene, toluene, ethylbenzene, and xylenes (BTEX) and semi-volatile organic compounds (SVOCs). To evaluate the effectiveness of the current MTBE treatment technologies, comprehensive treatability studies were conducted using groundwater samples collected from the JFK site. The treatability studies performed included sonochemical destruction of MTBE using ultrasound and O_3; advanced oxidation processes (AOPs) using O_3, O_3/UV, O_3/UV/photocatalyst (PC), and O_3/hydrogen peroxide (H_2O_2); and aerobic and anaerobic biodegradation.

This paper presents the results of the bench- and pilot-scale AOP treatability studies, and the results of the field implementation at the JFK site.

MATERIALS AND METHODS

Five 50-gallon drums of representative groundwater samples from NE-H10 area and three 50-gallon drums of representative groundwater samples from T8-G23 area of the JFK site were collected for the treatability studies.

The AOP reactors used for the design tests included the following:

AOP Reactor	Description
Glass Bubble Column	A "standard" apparatus used for conducting tests to compare with literature performance indices.
Mobile Test Unit (MTU)	A pilot test trailer equipped with advanced oxidation equipment at a scale that closely simulates a full-scale system.

Analytical Procedures. Except as otherwise noted, the analytical procedures followed the "Standard Methods for the Examination of Water and Wastewater, 19th Edition." Significant analytical procedures used in the test program include the following:

Analytical Parameter	Method	Detection Range/Limit
O_3 in water	Hach Method 8177 (Accuvac)	0.01 to 1.5 mg/L
O_3 in gas	IOA 001/87 (F)	100 to 1 g/m^3
H_2O_2 in water	Hach Test Kit Model HYP-1	0.2 to 10 mg/L
MTBE	GC/MS with Purge and Trap	5 µg/L
BTEX	GC/MS with Purge and Trap	0.5 µg/L
MTBE byproducts	GC/MS with Purge and Trap	5 µg/L
SVOCs	EPA 5030A/8270	Matrix Specific
Metals	Total Metals ICAP Scan SM18 3030 B/D & 3120 B	Matrix Specific
	Mercury by CVAA SM18 3112 B	
Bromide/Bromate	IC Anion Scan SM18 4110 B	0.1 to 1 mg/L

For trials with H_2O_2, the peroxide was neutralized with an excess of sodium thiosulfate and then analyzed for H_2O_2 using a Hach method.

RESULTS AND DISCUSSIONS

MTBE Destruction in a Glass Reactor. To evaluate the effectiveness of O_3/H_2O_2 treatment on MTBE, a bench-scale treatability study was performed at

Caltec Laboratory in Pasadena, California. Table 1 shows the characteristics of the groundwater samples from the JFK site. In a typical O_3/H_2O_2 experiment, O_3 gas was introduced through a disk type of gas diffuser installed at the bottom of a 2 L glass reactor into distilled water containing 0.7 mM of MTBE at 15°C. The gas flow rate was maintained at 240 mL min^{-1}. O_3 was dosed at a rate of 11.9 mg L^{-1} min^{-1}.

The O_3/H_2O_2 treatment process was found to be very effective for the destruction of MTBE and even more effective in bicarbonate-spiked water, as shown in Table 2. The higher rate of reaction in the O_3/H_2O_2 runs at higher bicarbonate levels suggests that the carbonate radical is highly reactive toward MTBE and that it functions as an effective ·OH radical trap that is able to engage in further oxidative reactions with MTBE. We found that MTBE degradation generally follows a pseudo first-order kinetic reaction.

The principal reaction intermediates that were observed in these reactions were tertiary butyl alcohol (TBA), methyl acetate, acetone, and tertiary-butyl-formate (TBF). Of these reaction intermediates, the TBF yield was the highest, but at the same time it also reacts with hydroxyl radical and is, therefore, completely destroyed after 15 minutes of treatment.

MTBE Destruction in a Glass Bubble Column Reactor. To further verify the effectiveness of O_3/H_2O_2 treatment on MTBE destruction, a similar treatability was performed at Hydroxyl System BC, Canada using a glass bubble column. Table 2 shows the oxidant dose rates and resulting MTBE degradation half-lives. Similar degradation trends were found in these experiments, as shown in Figure 1. These results prove that MTBE can be effectively destroyed with O_3/H_2O_2 treatment.

TABLE 1. JFK Groundwater Characteristics

Parameter	Unit	Concentration
Organics:		
MTBE	µg/L	73,000
Benzene	µg/L	4,700
Toluene	µg/L	1,100
Ethylbenzene	µg/L	<100
Xylenes	µg/L	370
Naphthalene	µg/L	30
General Inorganics:		
Total Alkalinity	mg/L	608
Carbonate	mg/L	<1
Bicarbonate	mg/L	608
BOD5	mg/L	15
COD	mg/L	1,990
TOC	mg/L	45
Metals:		
Calcium	mg/L	110
Magnesium	mg/L	35.8
Iron	mg/L	24.3

TABLE 2. O_3/H_2O_2 Treatment of MTBE at 15 °C.

Oxidant Dose Rate	HCO_3^-	DW k_o, s^{-1}	JFK-2 k_o, s^{-1}	Half-Life, min
O_3 (11.9 mg O_3/L-min)	No	0.57×10^{-4}	NR	203
$O_3 + H_2O_2$ (11.9 mg O_3/L-min) (5 mg H_2O_2/L-min)	No	26.7×10^{-4}	NR	4.3
$O_3 + H_2O_2$ (11.9 mg O_3/L-min) (5 mg H_2O_2/L-min)	Yes 8.9 mM	57.5×10^{-4}	NR	2.0
O_3 (4.3 mg/L-min)	Yes	NR	1.97×10^{-4}	59
O_3 (11.9 mg O_3/L-min)	Yes	NR	4.08×10^{-4}	28
$O_3 + H_2O_2$ (11.9 mg O_3/L-min) (2 mg H_2O_2/L-min)	Yes	NR	34.7×10^{-4}	3.3
$O_3 + H_2O_2$ (11.9 mg O_3/L-min) (4 mg H_2O_2/L-min)	Yes	NR	29.3×10^{-4}	3.9
$O_3 + H_2O_2$ (11.9 mg O_3/L-min) (6 mg H_2O_2/L-min)	Yes	NR	39.8×10^{-4}	2.9

Note: NR = Not Run; DW = Distilled Water; JFK-2 = JFK groundwater

MTBE AOP Treatment Pilot Test. An AOP pilot test was performed in a 10 gallons per minute (gpm) Mobile Test Unit (MTU) using the optimal operating conditions developed from the bench-scale treatability study (Figure 2). The MTU is capable of carrying out several different advanced oxidation processes, which include ozonation, O_3/H_2O_2, O3/UV, O_3/UV/PC, and H_2O_2/UV. Groundwater samples from the NE-H-10 and T8-G23 areas of the JFK site were tested. As demonstrated in Figure 3, MTBE can be effectively destroyed to below the detection limit (5 µg/L) in both JFK groundwater samples.

AOP Treatment Intermediates. The degradation intermediates of AOP treatment were identified as acetone, methyl acetate, and tert-butyl alcohol (TBA). However, acetone, TBA, and methyl acetate are the predominant byproducts in AOP degradation. Acetone is produced in the greatest quantity and is more persistent than any of the other intermediates. Further testing with O_3/H_2O_2, O_3/UV, or O_3/UV/PC showed that acetone was persistent after 120 minutes of treatment. A fluidized bed bioreactor (FBB) was, therefore, built to treat the intermediate byproducts from the AOPs. We found that the byproducts, including acetone, TBA, and methyl acetate, were biodegraded to below the detection limit (5 µg/L). In an effort to meet the National Pollutant Discharge Elimination System (NPDES) discharge limit for acetone (50 µg/L), an FBB system was added to the overall treatment system.

FIGURE 1. Ozone/Hydrogen Peroxide Treatment of MTBE Using a Glass Bubble Column Reactor.

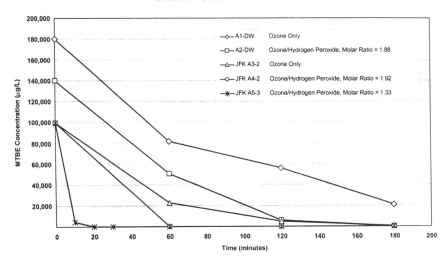

FIGURE 2. Mobile Test Unit (MTU).

FIGURE 3. Effects of Ozone/Hydrogen Peroxide on MTBE and Intermediates JFK H10 Water (Run 2E4).

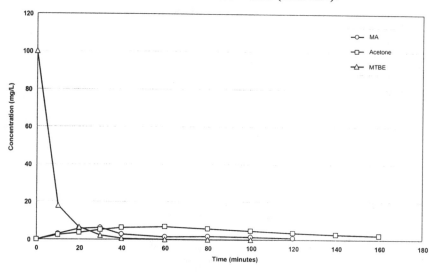

FIGURE 4. A Full-Size MTBE Treatment System.

Full-Scale System Implementation. Based on the design parameters gathered from the bench- and pilot-scale treatability studies, URS worked with Carbtrol Corporation and Hydroxyl System to design and build a 300 scfm multi-phase extraction and treatment system (MPE&TS) for light non-aqueous phase liquid (LNAPL) removal and a 20 gpm AOP/FBB groundwater treatment system (Figure 4) for the JFK site. The MPE&TS is used to recover free products from the site. Approximately 5,500 gallons of product have been recovered since the startup. The LNAPL removal and MTBE treatment systems were constructed in early 2000 and started up in June 2000. Figure 5 shows the performance of the AOP treatment system. The field results show that almost all the MTBE was destroyed in the AOP reactors with trace amounts going into the FBB. The effluent discharge limitations for MTBE, BTEX, and acetone are 50 µg/L. There is no current limitation on TBA. Figure 5 also demonstrates that the discharge limitations have been met since the startup.

FIGURE 5. Gate 23 Full Scale MTBE Treatment Performance.

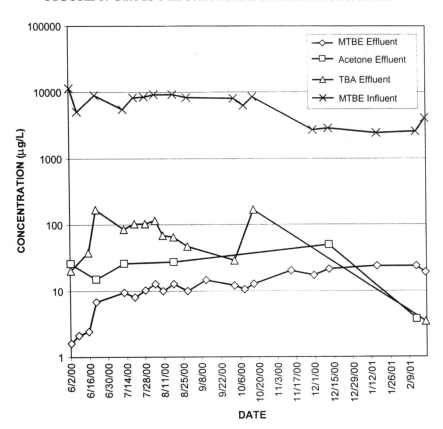

REFERENCES

Barker, J. F., C. E. Hubbard, and L. A. Lemon. 1990. pp. 113-127. In: *Proc. 1990 Petroleum Hydrocarbons and Organic Chemicals in Groundwater: Prevention, Detection, and Restoration.* Oct. 31 – Nov. 2. American Petroleum Institute and Association of Groundwater Scientists and Engineers: Houston, TX.

Mo, K., C. O. Lora, A. E. Wanken, M. Javamardian, X. Yang, and C. F. Kulpa. 1997. "Biodegradation of methyl *t*-butyl ether by pure bacterial cultures." *Appl. Microbiol. biotechnol.* 47:69-72.

Mormile, Mr. R., S. Liu, and J. M. Suflita. 1994. "Anaerobic biodegradation of gasoline oxygenates: extrapolation of information to multiple sites and redox conditions." *Environ. Sci. Technol.* 28:1727-1732.

Salanitro, J. P., L. A. Diaz, M. P. Williams, and H. L. Wisniewski. 1994. "Isolation of a bacterial culture that degrades MTBE." *Appl. Environ. Microbiol.* 60:2593-2596.

Suflita, J. M., and M. R. Mormile. 1993. "Anaerobic biodegradation of known and potential gasoline oxygenates in the terrestrial subsurface." *Environ. Sci. Technol.* 27:976-978.

Yeh, C. K., and J. T. Novak. 1994. "Anaerobic biodegradation of gasoline oxygenates in soil." *Water Environ. Res.*, 66:744-752.

COST OF MTBE REMEDIATION

Barbara H. Wilson, Hai Shen, and Dan Pope
(Dynamac Corp., Ada, Oklahoma 74820 USA)
Steve Schmelling (U.S. EPA, Ada, Oklahoma 74820 USA)

ABSTRACT
Widespread contamination of methyl *tert*-butyl ether (MTBE) in ground water has raised concerns about the increased cost of remediation of MTBE releases compared to BTEX-only sites. To evaluate these costs, cost information for 311 sites was furnished by U.S. EPA Office of Underground Storage Tanks (OUST, 2000), several states, BP/Amoco, Creek and Davidson (1998), and other sources. The majority of the sites were from South Carolina (183), Kansas (53), and New York (32). Information from sites in Maine, Texas, California, and Illinois was also included. The provided costs were project costs, actual costs, estimated project costs, or estimated project costs to date and included site assessment costs, capital expenditures, and operating and maintenance (O&M) expenses. BP/Amoco provided site specific cost data, average O&M costs, and total project costs for several sites primarily in California and on the East Coast.

The average cleanup cost of the 311 sites was $200,827 with a range of $21,000 (a UST in South Carolina) to $1,203,168 (for cleanup of a public water supply in Kansas), a standard deviation of $193,210, and a median of $150,000. The majority of the costs ranged from <$100,000 to $300,000 per site. The 311 sites included 183 sites from South Carolina that provided costs for remediation and operation and maintenance but no site assessment costs. The average cleanup cost for South Carolina was $146,132 compared to $279,022 for the remaining 128 sites.

Out of the 311 MTBE sites, 276 were service stations or other petroleum-related facilities, 32 of the sites were impacted drinking water wells/supplies, and three were hazardous waste sites. Overall, cleanup costs for MTBE contamination of drinking water wells were found to be higher than cleanup costs for USTs. The average cleanup cost of sites impacting drinking water supplies was $414,273 while the average cleanup cost for 276 service station/petroleum sites was $174,820. Excluding one major pipeline spill, the average cleanup costs for 275 service stations was $171,284. For many states, the presence of MTBE did not affect the cost of UST remediation or the technologies used for cleanup. The cost of cleanup of MTBE sites appeared controlled primarily by two factors: (1) MTBE contamination is the responsibility of most states' UST trust funds/programs, and (2) MTBE contamination impacting private or municipal drinking water wells substantially increases cleanup costs compared to cleanup costs for USTs that do not.

INTRODUCTION
Individual states vary in their responses to MTBE releases. Remediation decisions are frequently based on the location of a release in relation to a receptor.

Most MTBE sites not impacting drinking water wells are actively remediated by some combination of air sparging, soil vacuum extraction, and soil excavation. Bioremediation and natural attenuation are also widely used as remedies. Most MTBE sites impacting drinking water wells are remediated using multiple technologies, with pump-and-treat/air stripping/carbon treatment of the water prior to distribution. Increased costs are frequently due to provision of an alternate drinking water supply, and point-of-entry (home) treatment may also be used in some locations.

The majority of the MTBE sites are remediated under the various state UST programs. These programs provide funds for UST cleanups that must meet specific cost requirements. The types of programs include reimbursable, preapproval, and pay-for-performance. These cost guidelines include rates for labor, drilling, analytical costs, and any other potential costs associated with the UST cleanup. Most of the states have a deductible that must be met prior to the state issuing any funds. Basically in many states, the cost of MTBE cleanup will depend on the allowable cost expenditures for the individual states.

PLUME LENGTH COMPARISON OF MTBE AND BTEX

There is a general perception that the remediation requirements of MTBE and BTEX from fuel spills are different because MTBE is less biodegradable, because MTBE is less subject to natural attenuation, and because MTBE tends to form a larger plume than contamination from BTEX. Of concern to many regulators is the length of the MTBE plume compared to the BTEX plume and the potential for increased MTBE plume length to affect cleanup costs. The plume length of both MTBE and BTEX were compared in Table 1 for 215 South Carolina sites. Both active corrective action and monitored natural attenuation sites were compared for equal plume lengths, BTEX >MTBE, and MTBE >BTEX.

Out of the 215 sites, MTBE plume length was greater than the BTEX plume at 44 sites (20.7%). For the active corrective action sites with MTBE plume length greater than BTEX, the average MTBE plume length was 302 ft compared to 204 ft for the BTEX plumes. Of these sites, one MTBE plume was much longer at 1200 ft compared to 450 ft for the BTEX plume. At most sites, the BTEX plumes and the MTBE plumes overlap, and treatment for one has also been treatment for the other.

Because the MTBE plume is congruent to the BTEX plume, or is contained within the BTEX plume, MTBE is treated at most sites with technology that is also intended to treat BTEX contamination. As a consequence, MTBE is treated at most sites with technology that was originally developed for BTEX.

REMEDIAL TECHNOLOGIES USED AT USTs

The data provided by New York indicate that the use of a single technology for remediation of USTs is not common: normally two or three major technologies are implemented at each site. Table 2 compares the frequency of application of remedial technologies for cleanup of 1,563 UST releases in the State of New York. No cost information was provided for the majority of these sites. Most frequently, a suite of technologies was used a particular site. The most common approaches

TABLE 1. BTEX plume length compared to MTBE plume length at 215 South Carolina UST sites.

Comparison of plumes	Number of ACA Sites	Percent of sites	Number of MNA Sites	Percent of sites	Total Number of Sites	Percent of total sites
BTEX = MTBE	75	34.9	44	20.5	119	55.3
BTEX > MTBE	42	19.5	10	4.7	52	24.2
BTEX < MTBE	28	13.0	16	7.4	44	20.5
Total	145	67.4	70	32.6	215	100

ACA - Active corrective action
MNA - Monitored natural attenuation

were combinations of soil excavation, soil vacuum extraction (SVE), air sparging, and groundwater pumping. The highlighted cells in the rows indicate the number of times a particular technology was used for cleanup at the 1,563 UST sites. The values in the other columns designate the additional technologies used. For example, out of 1,563 sites, soil excavation was used at 601 sites. Out of these 601 sites, 206 sites also used SVE, 182 used pump-and-treat, 79 used air sparging (ground water), 106 used natural attenuation (ground water), 30 used other ground water treatments, 24 used dual phase extraction, 28 used O_2 injection (ground water), 28 used soil bioremediation, and 22 used other non-specified treatments. In the State of South Carolina, natural attenuation, air sparging and soil vacuum extraction were the most ordinarily used technologies for MTBE sites (data not shown). However, in the States of Texas and Tennessee, natural attenuation was the remedy chosen most frequently for USTs, followed by pump-and-treat and soil vacuum extraction (data not shown).

CLEANUP COSTS

To determine the effectiveness and cost of remedial actions currently used at leaking underground storage tank (LUST) sites, data from actual sites was compiled and analyzed. The 311 sites included 183 sites from South Carolina that provided costs for remediation and operation and maintenance but no site assessment costs; however, costs for the other states did include site assessment in addition to remediation, and operation and maintenance costs. Cost averages for both the 183 South Carolina sites and the 128 remaining sites were included in Table 3. The South Carolina sites were classified as service station/petroleum sites because no indication of impact to a drinking water well was found. The average cleanup cost for South Carolina was $146,132 compared to $279,022 for the remaining 128 sites. The average cleanup costs for the 93 service station/petroleum sites not in South Carolina were $231,270. This suggests that average site assessment costs may range from approximately $85,000 to $140,000.

TABLE 2. Cleanup technologies used for 1,563 USTs in the State of New York.

	Soil Excav.	SVE	GW P&T	GW Air Sparging	GW NA	GW Other Treat.	GW Dual Phase	GW O$_2$ Inject.	Soil Biorem.	Others
Soil Excav.	601	206	182	79	106	30	24	28	28	22
SVE		525	234	176	48	26	33	33	22	27
GW P&T			393	67	30	17	19	15	18	18
GW Air Sparging				186	10	4	7	10	8	10
GW NA					181	6	6	9	18	15
GW Other Treat						82	5	0	12	9
GW Dual Phase							76	3	2	3
GW O$_2$ Inject.								75	0	15
Soil Biorem.									60	13
Others										127

The provided cost data varied between the individual states that participated in the survey. New York provided cost data for 52 UST sites impacted with MTBE that had been closed between 7/98 and 7/00. These costs ranged from a minimum of $99 to $567,136, with 18 of the 52 sites <$20,000. No explanation was provided on how the costs were incurred. Including these low costs in their data resulted in the average cost of $97,399 to treat a MTBE site in New York compared to $200,827 for all sites in this report. New York may have included costs for sites that were evaluated as no-risk, whereas Kansas, South Carolina, and other states did not include no-risk costs. Cost data for New York ≤$20,000 were arbitrarily excluded from calculated averages presented in Table 3. A frequency distribution of cleanup costs for the 311 sites is presented in Figure 1.

COMPARISON OF COSTS FOR MTBE AND BTEX

A comparison of the mean cleanup costs calculated in this study with mean costs for the technologies of monitored natural attenuation (MNA), pump-and-treat, soil vacuum extraction/air sparging, soil vacuum extraction/pump-and-treat, and soil vacuum extraction/air sparging/pump-and-treat used in the States of Tennessee and Texas are presented in Table 4. The individual technology cost values were obtained from a nationwide site survey and from state UST programs as part of a critical review of the cost effectiveness of MNA in the risk management of petroleum hydrocarbon plumes (Chen and Fishman, 1999). In general, the cost of remediating a UST site with both MTBE and BTEX contamination ranges from <$100,000 to

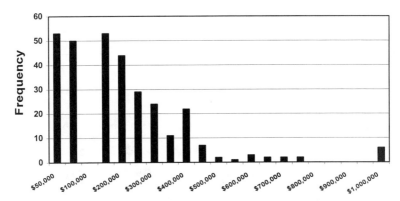

Figure 1. MTBE Cleanup Costs for 311 Sites

$300,000. However, if a private or municipal well is impacted or may potentially be impacted by MTBE, the average cost increases to approximately $415,000.

The U.S. EPA (OUST, 2000) conducted a survey of state regulators and regulators in the U.S. EPA regions to gather information on the relative costs of cleanup of MTBE and BTEX in their states. The U.S. EPA OUST, and the States of New York, Maine, Illinois, Ohio, Florida, Kansas, and South Carolina provided general information about the cost of UST site cleanups both with and without MTBE and about actions taken at UST release sites. The following summarizes their responses. The average costs to cleanup sites varied greatly from $750,000 in Illinois to $42,000 in Maine. In the State of Illinois, the average cost to successfully remediate sites with MTBE was estimated to be $750,000; the average cost to remediate sites without MTBE was estimated to be $95,000. Illinois expects 90% of the MTBE groundwater contamination sites to significantly add costs to the total cleanup, depending on the groundwater receptor impacts, because most of these sites have impacted high capacity drinking water wells. The State of Ohio estimates remediation of MTBE sites to be approximately $200,000, assuming a drinking water receptor is or may become impacted. The estimated average cost to remediate a BTEX-only groundwater contamination site in Ohio is about $130,000. The historical average to cleanup a UST site in Florida is about $250,000. It is estimated that in the State of Florida, the presence of MTBE will cause a 1% to 2% increase in cleanup cost compared to sites without MTBE. The State of South Carolina did not find a difference in treatment costs between MTBE and BTEX-only sites, with an average of $140,000 for each type of site. The average cost in the State of Maine to remediate UST sites both with and without MTBE is approximately $42,000, although costs associated with MTBE impact of drinking water supplies ranged from $200,000 to >$1,000,000. New York also observed a cost difference with an average of $97,400 for MTBE sites and $56,393 for BTEX-only sites.

To determine if any relationship could be identified between cleanup costs and site or plume characteristics, the MTBE cleanup costs for the South Carolina

TABLE 3. Total project cost by type of site.

Type of Site	No. of sites	Mean cost	Median cost	Std. Dev.	Minimum cost	Maximum cost
Total Sites: 311						
Drinking water supply	32	$414,273	$262,550	$379,490	$31,004	$1,203,168
Service station/ Petroleum	276	$174,820	$139,790	$139,890	$21,000	$1,147,000
Hazardous waste	3	$316,667	$400,000	$144,337	$150,000	$400,000
Total sites	311	$200,827	$150,00	$193,210	$21,000	$1,203,168
Sites Outside South Carolina: 128						
Drinking water supply	32	$414,273	$262,550	$379,490	$31,004	$1,203,168
Service station/ Petroleum	93	$231,270	$195,158	$173,506	$21,369	$1,147,000
Hazardous waste	3	$316,667	$400,000	$144,337	$150,000	$400,000
Total non-SC sites	128	$279,022	$207,000	$252,192	$21,369	$1,203,168
South Carolina Sites**: 183						
Service station/ Petroleum	183	$146,132	$117,000	$109,066	$21,000	$454,000

**South Carolina cost data are for remediation and O&M; no costs for site assessment were provided. Cost data for states other than South Carolina included site assessment, remediation, and O&M costs.

sites were plotted against BTEX soil concentrations, BTEX groundwater concentrations, MTBE groundwater concentrations, BTEX plume length, MTBE plume length, BTEX plume volume, and MTBE plume volume (data not shown). No clear correlation between cost of cleanup and any plume characteristic could be identified. For the South Carolina data, the cost associated with cleanup of MTBE sites could not be predicted based on size of release, length of plume, concentration of MTBE or BTEX, or any other measurable plume characteristic.

SUMMARY

There are several overall generalizations that may be made from the data presented in this paper. In general, the average cost to remediate a MTBE site with drinking water impacts is approximately twice the cost of cleanup of sites where there are no drinking water impacts ($414, 273 compared to $174,820). The cleanup technologies used most often in the states surveyed were soil excavation, soil vacuum extraction, pump-and-treat, air sparging, and natural attenuation, with multiple technologies most frequently applied for site remediation. It is difficult to estimate remedial cost based on quantifiable characteristics of the release, the resulting plume(s), the contaminant type, or the contaminant concentration.

TABLE 4. Comparison of mean costs for total project cost and for several technologies.

Source	MNA	P&T	SVE/AS	SVE/P&T	SVE/AS/ P&T	Total Project Cost
Tennessee UST*	$74,000	$122,209		$226,366	$328,017	
Texas UST*	$113,574	$364,700	$179,429	$265,948	$316,343	
This study: South Carolina			$176,556		$270,141	$146,132
This study: Creek and Davidson						$321,590
This study: UST						$171,284
This study: Water supply						$414,273
This study: All sites						$200,827

*Chen and Fishman, 1999.

There are individual responses of states to MTBE releases, and remediation decisions are frequently based on the location of a release in relation to a receptor. The approach that many states (other than California) take to the cleanup of USTs may be reflected in the following statement from Greg Hattan, KDHE, "The philosophy of our remedial effort at all sites is an aggressive source removal with treatment to the receptor, if it becomes impacted. We rarely address the plume other than AS in the hottest areas. We have found this to be efficient and cost effective. We have and do use risk based decision making (based on proximity to receptor) in determining which sites go into remediation."

DISCLAIMER

The views expressed in these Proceedings are those of the individual authors and do not necessarily reflect the views and policies of the U.S. Environmental Protection Agency (U.S. EPA). Scientists in the U.S. EPA's Office of Research and Development have prepared the U.S. EPA sections, and these sections have been reviewed in accordance with U.S. EPA's peer review and administrative review policies and approved for presentation and publication.

ACKNOWLEDGMENT

This paper would not have been possible without the assistance and information provided by the following agencies and individuals: Doug Maddox and Steve McNeely, U.S. EPA OUST; Art Shrader, South Carolina DHEC; Bill Reetz and Greg Hattan, Kansas KDHE; Tom Conrardy, Florida DEP; Bruce Hunter, Maine DEP; D. Darmer, New York NY DEC; Gilberto Alvarez, U.S. EPA Region 5 UST Program; Linda Fiedler, U.S. EPA, DC; and Tetra Tech EM Inc., Reston VA.

REFERENCES

Chen, J.S. and M. Fishman. 1999. *Critical Review: Cost-Effectiveness Analysis of Natural Attenuation in the Risk Management of Petroleum Hydrocarbon Plumes.* Prepared by Dynamac Corporation for U.S. EPA Subsurface Protection and Remediation Division, National Risk Management Research Center, Ada, OK.

Creek, D.N, and J.M. Davidson. 1998. "The Performance and Cost of MTBE Remediation Technologies." *Proceedings of the 1998 Petroleum Hydrocarbons and Organic Chemicals in Groundwater.* November 11-13, 1998. Houston, TX.

Hattan, G. 2000. KDHE. Personal communication.

Kolhatkar, R. 2000. Personal communication.

OUST. 2000. Office of Underground Storage Tanks. U.S. Environmental Protection Agency. Washington, DC. Doug Maddox and Steve McNeely. OUST State Survey: State of South Carolina: Art Shrader, DHEC. State of Kansas: Bill Reetz and Greg Hattan, KDHE. State of Florida: Tom Conrardy DEP. State of Maine: Bruce Hunter, Maine DEP. State of New York: D. Darmer, NY DEC. Information concerning Ohio and Illinois was provided by Gilberto Alvarez, U.S. EPA, Region 5 UST Program. Some information was provided to OUST by Linda Fiedler, U.S. EPA, DC through Tetra Tech EM Inc., Reston VA.

BIOREMEDIATION OF MTBE THROUGH AEROBIC BIODEGRADATION AND COMETABOLISM

Kristen E. Hartzell, Victor S. Magar, James T. Gibbs, Eric A Foote, Christy D. Burton
(Battelle, Columbus, Ohio)

ABSTRACT: A laboratory-scale microcosm study was conducted to determine whether site-specific soil microbes could biodegrade methyl *tert*-butyl ether (MTBE) under anaerobic, aerobic, or cometabolic conditions. Microcosms were constructed using soils and groundwater from both the Department of Defense Housing Facility (DoDHF) Novato, California, and the naval Base Ventura County Port Hueneme (NBVC), California. Aerobic and cometabolic conditions were investigated with and without nitrate addition. Cometabolism was investigated using either propane or butane gas as a primary growth substrate. The laboratory results indicated that MTBE could be biodegraded under aerobic conditions at each site. Overall, cometabolic growth substrates did not enhance or stimulate MTBE biodegradation. Nitrate was required for cometabolic degradation, and anaerobic biodegradation was not observed. In situ bioremediation through aerobic air sparging appears to be a potential remediation treatment for consideration at the DoDHF Novato and NBVC PH sites.

INTRODUCTION

MTBE, a gasoline oxygenate additive, comprises 11% of Reformulated Gasoline (RfG) in California, and has been observed in groundwater at NBVC PH and DoDHF Novato. Due to the widespread use of MTBE since the late 1970s, it is commonly found in groundwater aquifers as a result of releases from leaking underground storage tanks. MTBE's high solubility (approximately 48 g/L) and low soil-water partitioning coefficient results in its affinity toward water, thereby causing transport of the compound in groundwater to greater extents than other gasoline constituents (e.g., benzene and toluene). The downgradient extent of MTBE dissolved in groundwater at NBVC PH has expanded approximately 1500 m (~4,900 ft) since 1985 (Salinitro et al. 2000). Site characterization of activities performed at DoDHF Novato indicated that the MTBE dissolved in groundwater extends approximately 880 m (~2,900 ft) from its original source area (Battelle, 2001).

Studies have shown that mixed microbial consortia can be used to successfully degrade MTBE (Salinitro et al. 1994, Eweis et al. 1997). For example, in one study, three propane-oxidizing bacterial strains were isolated and tested. Each strain was found to degrade MTBE after growth on propane (Steffan et al. 1997).

In order to predict future plume behavior or evaluate potential remedial alternatives for most MTBE-impacted sites, it is important to determine whether naturally occurring microbes are capable of aerobic or cometabolic MTBE biodegradation. The objective of this microcosm study was to assess the potential for aerobic and anaerobic MTBE biodegradation at two sites that have MTBE dissolved in groundwater, and to determine whether propane or butane can stimulate cometabolic MTBE degradation.

MATERIALS AND METHODS

Laboratory microcosms were prepared in 20-mL glass serum bottles capped with Teflon-lined butyl rubber stoppers secured with crimp seals. Microcosms were constructed using saturated soils and groundwater from both NBVC PH and DoDHF Novato. Two saturated soil and groundwater types were collected from the DoDHF Novato site. One set was collected from within the MTBE-only plume and the second set was collected from within the BTEX and MTBE plume, the latter will be called commingled Novato. Each microcosm contained approximately 5 g of soil from the specific site, and 10 mL of groundwater. The initial liquid MTBE concentration in the bottles was adjusted to approximately 10 mg/L. Aerobic and cometabolic mechanisms (propane and butane as growth substrates) were investigated both with (amended) and without (unamended) nitrate addition. The initial liquid nitrate concentration was adjusted to 30 mg/L. Killed controls were prepared using all soil types and 20 mg/L mercury dichloride ($HgCl_2$). Microcosms were prepared in batches and sacrificed in triplicate at each sampling event.

Aerobic conditions were established by initially displacing the headspace of each vial with ultra high purity (UHP) 50:50 $O_2:N_2$; the 50:50 mixture was used instead of UHP air to ensure sufficient oxygen supply throughout the experimental incubation period. The headspaces of cometabolic microcosms were flooded with 4% propane or butane blended with the UHP 50:50 $O_2:N_2$ mixture. Propane, butane and oxygen concentrations were verified independently using additional bottles that were immediately sacrificed during experimental setup. All gases were filter-sterilized at 0.25 µm. Anaerobic microcosms were constructed in a glove box, under a nitrogen/5% hydrogen atmosphere. Killed controls were prepared using 20mg/L $HgCl_2$. All microcosms were sealed with butyl-septa crimp caps.

Vials were stored at 18°C on shaker tables operated at 100 RPM. Triplicate microcosms from each condition were sacrificed at selected time intervals and analyzed for MTBE, TBA, TBF, acetone, 2-propanol, propane, butane, and nitrate. When MTBE was no longer detected in the triplicate bottles, the remaining equivalent bottles were respiked with MTBE, headspaces were reflooded with the respective gas mixtures, and the bottles were resealed. Microcosms that showed complete removal of MTBE after respiking underwent the same respiking process again. During the second respiking event, the headspace was flooded with UHP air (i.e., 20% O_2); cometabolic microcosms were maintained with initial 4% butane or propane concentrations blended with the UHP air.

Propane, butane, MTBE, TBA, and TBF, acetone, and 2-propanol, were analyzed by using a Teckmar 7000 headspace autosampler attached to a Varian Star 3400 gas chromatograph equipped with a flame ionization detector (FID). Nitrate was measured from the liquid phase, after the headspace autosampler analysis, using an ion chromatograph. The estimated method detection limits for; MTBE, TBA, TBF, acetone and 2-propanol were 82 µg/L, 158 µg/L, 216 µg/L, 134 µg/L, and 98 µg/L, respectively. Headspace O_2 was measured in separate sacrificed bottles using an SRI 8610A gas chromatograph equipped with a thermal conductivity detector (TCD). Headspace O_2 was measured only at the end of each MTBE spiking event to ensure that O_2 was not limiting in the bottles.

RESULTS

Aerobic. Port Hueneme (PH) and MTBE-only Novato aerobic microcosms showed complete MTBE removal after 65 days, which was confirmed with repeated spiking. Nitrate addition did not enhance MTBE degradation in the PH and MTBE-only Novato microcosms (Figure 1). The commingled Novato microcosms amended with nitrate showed complete MTBE removal after 120 days. The unamended commingled Novato microcosms showed no significant MTBE degradation (Figure 2). TBA and TBF were consistently low, and acetone and 2-propanol concentrations were consistently below detection limits. These results suggest complete MTBE mineralization under aerobic conditions, or transformation to an unknown byproduct. Both NBVC PH and DoDHF MTBE only Novato microcosms showed lag times of approximately 50 days. In addition, microcosms from both sites showed similar behavior after respiking.

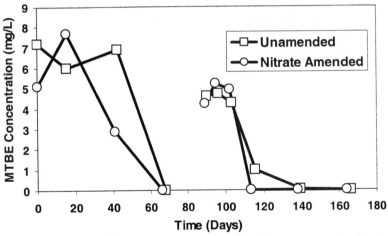

FIGURE 1. Novato aerobic microcosm results from the MTBE-only plume. MTBE was respiked into the microcosms at Day 90. Values are the average of triplicate samples.

Anaerobic. Anaerobic microcosms showed no sign of MTBE removal and no TBA or TBF accumulation during the 150-day incubation period (Figure 2).

Propane. All microcosms with nitrate added showed complete MTBE and propane removal, which was confirmed with repeated spiking (Figure 3).

Propane and nitrate addition enhanced MTBE degradation in the commingled Novato microcosms, but there was no benefit to propane addition in the Port Hueneme and MTBE-only Novato microcosms. Bottles with no added nitrate showed slower propane and MTBE removals. MTBE degradation was inhibited in bottles with high residual propane concentrations (Figure 4). MTBE degradation in the Port Hueneme microcosms was similar to degradation in the MTBE-only Novato microcosms.

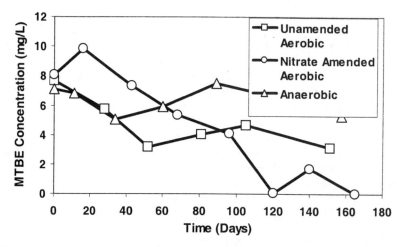

FIGURE 2. Novato aerobic microcosm results from the commingled MTBE plus TPH plume and anaerobic microcosm results from the MTBE-only plume. Values are the average of triplicate samples.

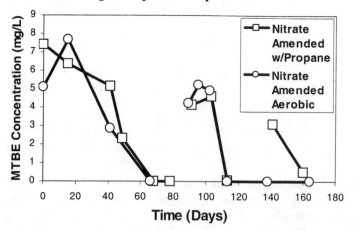

FIGURE 3. Novato aerobic and propane cometabolic microcosm results from the MTBE-only plume; nitrate was added to each microcosm. Propane degradation closely resembled MTBE degradation patterns. MTBE degradation was not enhanced by propane addition. Values are the average of triplicate samples.

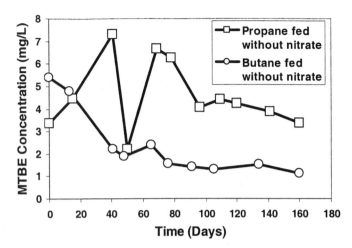

FIGURE 4. Novato aerobic propane and butane cometabolic microcosm results from the MTBE-only plume. Nitrate was not added. Degradation of propane and butane was negligible (data not shown), resulting in inhibited MTBE degradation. Values are the average of triplicate samples.

Butane. Port Hueneme and MTBE-only Novato microcosms with added nitrate showed complete butane and MTBE removal confirmed with repeated spiking (Figure 5). MTBE degradation was not enhanced by butane and nitrate addition. Nitrate was required for butane degradation. MTBE degradation was inhibited in the commingled Novato microcosms that had high residual butane concentrations (Figure 4). Significant TBF concentrations (2mg/L) were detected in unamended MTBE-only Novato microcosms with butane.

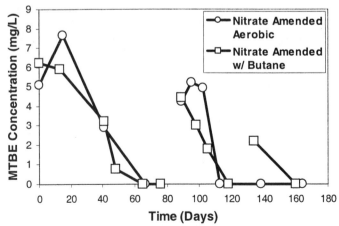

FIGURE 5. Novato aerobic and butane cometabolic microcosm results from the MTBE-only plume; nitrate was added to each microcosm. MTBE degradation was not enhanced by butane addition. Values are the average of triplicate samples

The results of this study demonstrated aerobic biodegradation potential at both sites. These results are consistent with the findings of J. Salanitro, et al., who reported that in situ MTBE biodegradation is enhanced through O_2 sparging at Port Hueneme (Salinitro, J.P., et al. 2000). Cometabolic microcosms with nitrate showed complete MTBE removals with propane or butane removals. However, there was no apparent advantage to MTBE degradation conferred by the cometabolic microcosms compared to the aerobic microcosms, with the exception of the commingled Novato microcosms.

The confirmation of aerobic biodegradation capacity at these sites indicates that the list of potentially applicable alternatives for plume restoration and intraplume migration control can be expanded to include in situ approaches such as biobarriers, biosparging (possibly without SVE), and oxygen releasing compounds. On site aerobic bioreactors can be considered for treating extracted water in the event that a groundwater extraction alternative is selected for plume restoration or control.

This information further reduces uncertainty about the fate and transport of MTBE and the likely future behavior of the MTBE dissolved in groundwater at the sites addressed in this study. For example, at DoDHF Novato, decreasing estimates of total dissolved MTBE and decreasing maximum and average concentrations are consistent with aerobic biodegradation supported by the infiltration of oxygen-rich rainwater. The absence of observable anaerobic biodegradation is consistent with the distribution of elevated MTBE concentrations in groundwater areas containing little or no dissolved oxygen at the Novato site. More field observations are necessary before making these types of comparisons at the NBVC PH site.

Comparison of aerobic and cometabolic biodegradation results calls into question the advantage offered by injecting propane or butane for cometabolism at these sites. The results indicate that it would be unlikely to recoup costs expended on propane purchasing, handling, and specialized design, and that simple in situ oxygenation is likely to enhance biodegradation. Pilot systems should be demonstrated at the site(s) to confirm that the laboratory results produced from this work are reproducible in situ.

REFERENCES

Salinitro, J.P., P. C. Johnson, G. E. Spinnler, P. M. Maner, H. L. Wisniewski, and C. Bruce. 2000. "Field-Scale Demonstration of Enhanced MTBE Bioremediation through Aquifer Bioaugmentation and Oxygenation." *Environ. Sci. Technol.* 34. 4152-4162.

Battelle. 2001. *Final Remedial Investigation Report for Former UST Site 957/970 at Department of Defense Housing Facility (DoDHF) Novato, California.* Prepared for the Southwest Division, Naval Facilities Engineering Command under NFESC Contract No. N47408-95-D-0730. January 31.

Salinitro, J.P., L.A. Diaz, M.P. Williams, and H.L. Wisniewski. 1994. "Isolation of a bacterial culture that degrades methyl *t*-butyl ether." *Appl. Environ. Microbiol.* 60:2593-2596.

Eweis, J.B., D.P. Chang, E.D. Schroeder, K.M. Scow, R.L. Morton, and R.C. Caballero. 1997. Presented at the Air and Waste Management Association 90th Annual Meeting and Exhibition, Toronto, Ontario, Canada. AWMA, Washington D.C.

Steffan, R.J., K. McClay, S. Vainberg, C.W. Condee, and D. Zhang. 1997. "Biodegradation of the Gasoline Oxygenates Methyl tert-Butyl Ether, Ethyl tert-Butyl Ether, and tert-Amyl Methyl Ether by Propane-Oxidizing Bacteria." *Appl. Environ. Microbiol.* 63. 4216-4222.

COMETABOLISM OF MTBE BY AN AROMATIC HYDROCARBON-OXIDIZING BACTERIUM

Michael Hyman (North Carolina State University, Raleigh, NC)
Christy Smith (North Carolina State University, Raleigh, NC)
Kirk O'Reilly (Chevron Research and Technology Company, Richmond, CA)

ABSTRACT: A novel bacterium, strain VB-1, was isolated from propane-enrichment culture seeded with MTBE-contaminated sediment from Vandenburg AFB. This organism is able to cometabolically degrade MTBE after growth on a range of linear and branched alkanes (C_3-C_8) and aromatics (benzene and toluene). Oxidation of MTBE by cells grown on each of these substrates results in the production of both *tertiary* butyl formate (TBF) and *tertiary* butyl alcohol (TBA). Cells grown on each of these substrates also oxidize TBA and the oxidation of both MTBE and TBA is inhibited in all instances by the presence of acetylene. The kinetics of MTBE oxidation suggest that this organism expresses at least two different enzymes that are capable of oxidizing MTBE. Our results also suggest that propane is a substrate for the enzyme responsible for initiating the oxidation of aromatics. The significance of our findings is that organisms have now been identified that can cometabolically degrade MTBE after growth on most of the major components of gasoline.

INTRODUCTION

Methyl *tert*-butyl ether (MTBE) is added to many modern gasoline formulations as both an octane enhancer and as an oxygenate. Several bacteria have been isolated that can grow on MTBE. However, the majority of bacteria that have been described that can degrade MTBE do so through fortuitous cometabolic processes after growth on another substrate. Our recent research has demonstrated that organisms that can grow on simple branched alkanes such as isopentane consistently cometabolically degrade MTBE (Hyman et al., 2000). Other studies have also demonstrated that compounds such as n-pentane and cyclohexane can also stimulate cometabolic MTBE degradation in both pure and mixed cultures. A common theme that ties these recent reports together is that they demonstrate that many of the major components of gasoline can apparently serve as stimulants for MTBE cometabolism. As the majority of MTBE is introduced into the environment as part of gasoline, the biodegradation of gasoline hydrocarbons is likely to have a large impact the environmental fate of MTBE and vice versa. In this study we describe the activities of a bacterium that was recently isolated from MTBE-impacted sediments from Vandenburg AFB. This organism, strain VB-1, was isolated from an enrichment culture established using propane as a sole source of carbon and energy. The isolated organism is able to cometabolically degrade MTBE after growth on many alkane components of gasoline as well as benzene and toluene.

MATERIALS AND METHODS

Enrichment cultures were established by inoculating sediment samples (~5 g) into a sterile centrifuge tube (50 ml) containing double distilled H_2O (45 ml). The tube was agitated for 30 s intervals (5 repetitions) using a sonic water bath. The agitated material was then allowed to settle for 30 min. Samples (0.5 ml) of the supernatant were then transferred into glass serum vials (100ml) that contained a mineral salts medium (30 ml). The vials were sealed with butyl rubber stoppers and crimp seals. Filter-sterilized propane (35 ml) was then added to the vials as an overpressure. The enrichment cultures were then incubated at 30° C for 3 weeks in a shaking incubator (150 rpm). A sample (0.15 ml) from the initial enrichment was then transferred into another sealed vial containing fresh media and propane as the sole source of carbon and energy. After further incubation for 2-3 weeks, samples (100 µl) of these cultures were spread on mineral salts agar plates. The inoculated plates were then incubated in a glass dessicator jar with propane (~10% gas phase) as the sole source of carbon and energy. Individual colonies were subsequently picked from these plates and were streaked out on mineral salts agar plates and reincubated in the presence of propane. The organism described in this report was found in high abundance and was chosen for further characterization. The morphological and structural characteristics of isolate were determined by standard techniques. 16S rRNA genes were amplified and isolated using universal primers 515F and 1492R (Hyman et al., 2000).

Larger scale liquid cultures of each isolate were grown at 30° C in minerals salts medium (100 ml) in glass bottles (600 ml) sealed with screw cap tops fitted with butyl rubber septa. All gaseous hydrocarbon growth substrates (60 ml) were added to the bottles as an overpressure. All liquid hydrocarbon growth substrates were added to an initial concentration of 0.05% vol/vol. After incubation for 5 days the bottles were opened under sterile conditions to replenish oxygen. The vials were then sealed again and the corresponding hydrocarbon was added to the culture at the concentrations described above. After an additional 24 h incubation the cells were harvested by centrifugation (10,000 x g for 10 min). The cell pellet was resuspended in sodium phosphate buffer (50 mM, pH 7.0) and was centrifuged again. The supernatant was discarded and the cell pellet was finally resuspended with buffer at a final protein concentration of ~ 10 mg protein ml^{-1}. These cell suspensions were stored on ice and were used in experiments within 4 h after harvesting.

The oxidation on MTBE by the isolated bacterium was investigated using small scale reactions maintained at 30° C in a shaking water bath (150 rpm). The reactions were conducted in glass serum vials (10ml) stoppered with butyl rubber septa and aluminum crimp seals. The reaction vials contained buffer (~900µl) and varying quantities of MTBE-saturated buffer (0-15 µl [0-8.2 µmoles]). The reactions were initiated by the addition of cells (0.2 to 1.2 mg total protein) to the vials. Samples (2 µl) were removed from the reaction and were directly injected into a gas chromatograph equipped with a flame ionization detector. The reactants and products were separated using a 6 ft stainless column packed with Porapak Q (60-80 mesh) that was operated at 160° C. Nitrogen was used as carrier gas at a flow rate of 20 ml/min.

RESULTS

The isolated bacterium, strain VB-1, was identified as *Variovorax* species based on the 16SrRNA sequence obtained during this study. In addition to growth on the enrichment substrate, propane, this strain also grew well on a wide range of linear and simple branched alkanes. These substrates include n-butane, n-pentane, n-hexane and isopentane (Fig. 1). Slightly lower levels of growth were obtained using n-octane or isobutane as growth substrates.

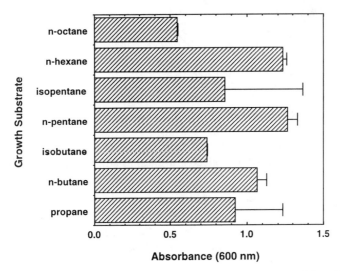

FIGURE 1. Growth of Strain VB-1 on n-alkanes and isoalkanes under standardized conditions (5 days at 30° C)

The ability of cells of strain VB-1 to cometabolically oxidize MTBE was determined using samples of concentrated cell suspensions (~ 1mg protein) incubated in short term assays (1h) in the presence of 1 µmole MTBE. Cells of strain VB-1 were able to rapidly degrade MTBE after growth on all of alkane substrates examined in the experiment described in Fig.1. In all cases, MTBE degradation was associated with the accumulation of two characteristic products of MTBE oxidation; *tertiary* butyl formate (TBF) and *tertiary* butyl alcohol (TBA) (Fig. 2). Oxidation of MTBE and the production of both TBF and TBA by cells grown on each alkane was also inhibited by acetylene (10% v/v gas phase). Acetylene is a potent and selective inhibitor of many monooxygenase enzymes. The effects of acetylene suggest that an alkane monooxygenase is responsible for the oxidation of MTBE in alkane-grown cells of strain VB-1.

During this study we also investigated whether strain VB-1 was able to utilize aromatic hydrocarbons as growth substrates. The bacterium grew consistently on both benzene and toluene. No growth was observed in the presence of other BTEX compounds over a period of 11 days (Table 1).

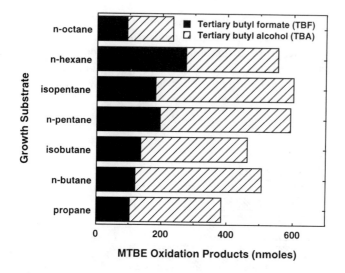

FIGURE 2: Production of TBF and TBA from MTBE (1 μmole) by cells of Strain VB-1 after growth on selected n-alkanes and isoalkanes

TABLE 1: Growth of Strain VB-1 on aromatic hydrocarbons

	Potential Growth Substrate					
	toluene	benzene	ethylbenzene	o-xylene	m-xylene	p-xylene
Starting OD	0.014	0.014	0.014	0.014	0.014	0.014
Ending OD	0.320	0.511	0.006	0.004	0.002	0.003

The MTBE-oxidizing activity of toluene- and benzene-grown cells was subsequently determined in similar experiments to those described above in Fig. 2. Cells grown on both aromatic hydrocarbons were able to rapidly oxidize MTBE and in both cases TBA and TBF were observed as oxidation products. The kinetics of MTBE oxidation by propane, toluene and benzene-grown cells of strain VB-1 were examined to provide a quantitative comparison of the MTBE-degrading activity of each cell type. A series of short-term incubations (30 min) were conducted using concentrated cell suspensions (~ 1mg total protein) incubated in the presence of a range of initial MTBE concentrations. The concentration of TBF and TBA detected in the reaction medium was then used to estimate the initial rate of MTBE oxidation.

Our results (Fig. 3) suggest that strain VB-1 has at least two different enzyme systems that are capable of degrading MTBE. The putative alkane monooxygenase activity observed in propane- and other alkane-grown cells has an estimated K_s value for MTBE of ~450 μM. In contrast, the estimated K_s values derived for the MTBE-oxidizing activity of benzene and toluene-grown are both

2-fold lower (270 µM) than that estimated for propane-grown cells. The similar K_s values for MTBE oxidation for benzene- and toluene-grown cells suggests that the same enzyme may initiate the oxidation of both aromatic substrates.

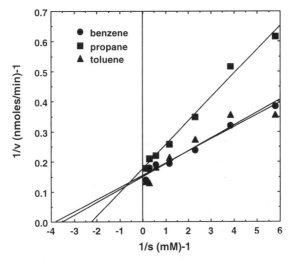

FIGURE 3: Kinetics of MTBE oxidation by propane, toluene and benzene-grown cells of Strain VB-1

The estimated K_s values provide some indication of differences between the enzyme systems expressed by cells of strain VB-1. However, these constants are derived by quantifying TBA and TBF accumulation and may be misleading because of the potential for the further oxidation of TBA (Hyman et al., 1998). To investigate this possibility, propane-, benzene- and toluene-grown cells of strain VB-1 were incubated with a fixed concentration of TBA (1 mM) in reactions conducted either in the presence or absence of acetylene (10% v/v gas phase). In each case the cells were able to consume TBA at substantial rates in the absence of acetylene and in all cases the consumption of TBA was inhibited by the presence of acetylene (Fig. 4). We conclude from these experiments that like MTBE, TBA is also a substrate for the oxygenase enzymes expressed by strain VB-1 after growth on alkanes or aromatics. The further oxidation of TBA has previously been observed in our studies of the oxidation of MTBE by a variety of alkane-oxidizing bacteria (Hyman et al. 1998, 2000).

In addition to oxidizing TBA, we observed that the MTBE-oxidizing activity of benzene-grown cells of strain VB-1 was also inhibited by the presence of alkanes. The presence of as little as 1% (v/v gas phase) propane inhibited MTBE degradation by benzene-grown cells by approximately 15% (Fig. 5). Higher levels of inhibition were observed as the propane concentration was increased and almost complete inhibition of MTBE oxidation was achieved with 10% (v/v gas phase) propane. The concentration dependence of this inhibitory effect suggested that the inhibition might arise due to competitive interactions between MTBE and propane for binding and oxidation by the oxygenase enzyme presumed to be responsible for initiating benzene oxidation. Further evidence to support this was obtained by analyzing the reaction medium for the appearance of distinctive oxidation products derived from propane. Both 1-propanol and 2-propanol accumulated in the reaction medium during the time course of the experiment. (Fig. 6). Furthermore, the rate at which these products accumulated was proportional to the concentration of propane added to the incubation.

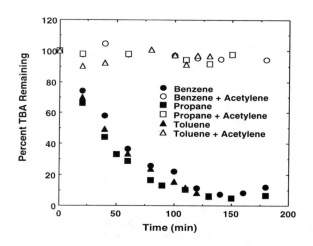

FIGURE 4: Acetylene inhibition of TBA consumption by propane-, toluene- and benzene-grown Strain VB-1 in the presence of 1 mM TBA

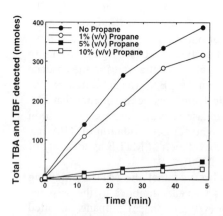

FIGURE 5: Effect of propane on the oxidation of MTBE by benzene-grown Strain VB-1

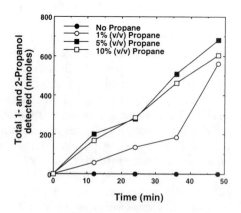

FIGURE 6: 1- and 2-propanol accumulation during propane inhibition of MTBE oxidation by benzene-grown Strain VB-1

DISCUSSION

The results we have obtained during this study focus on two physiological features of strain VB-1; the ability of this organism to degrade MTBE after growth on alkanes and the equivalent activity after growth on selected aromatics. The MTBE-oxidizing activity of strain VB-1 after growth on alkanes is very similar to the activities we have previously described for organisms originally isolated on isobutane (Hyman *et al.*, 2000) and other alkanes (Hyman *et al.*, 1998) and well-characterized organisms such as *Mycobacterium vaccae* JOB5. The similarities include the fact that strain VB-1 can oxidize MTBE after growth on a range of linear and branched alkanes (Fig. 1) and that both TBA and TBF are generated as immediate MTBE oxidation products. Further similarities include the fact that cells grown on propane can also oxidize TBA, and the fact that the oxidation of both MTBE and TBA is inhibited by the oxygenase inhibitor, acetylene. Although further research is obviously needed, we consider it very likely that strain VB-1 expresses the same oxygenase enzyme in response to all of the alkane substrates we have examined in this study. The similar K_s value we have determined for MTBE oxidation by propane-grown cells of strain VB-1 (Fig. 3) and other alkane-utilizers also suggests that the putative oxygenase enzymes involved in both alkane and MTBE oxidation in all of these organisms may also be very similar at the molecular level.

The second physiological feature of strain VB-1 is its ability to oxidize MTBE after growth on selected aromatics. This is a novel and potentially more significant activity than the activity observed after growth on alkanes. Our kinetic analysis of MTBE oxidation (Fig. 3) suggests that strain VB-1 expresses a different oxygenase after growth on aromatics as opposed to alkanes. However, the effects of acetylene on MTBE and TBA oxidation and the production of TBA and TBF from MTBE by cells of strain VB-1 grown on aromatics are all features that suggest a monooxygenase enzyme is involved in initiating the oxidation of both MTBE and aromatics in this organism. If these conclusions are correct, one wonders why MTBE-degrading activity is not observed in bacteria that are known to express monooxygenase enzymes to initiate the oxidation of aromatics (Hyman *et al.*, 1999)? One possible explanation is that none of aromatic hydrocarbon-oxidizers we have previously tested for MTBE-degrading activity are known to be able to oxidize alkanes after growth on aromatics. The fact that benzene-grown cells of strain VB-1 can apparently oxidize propane to a mixture of 1- and 2-propanol (Fig. 5 & 6) may point to a novel class of monooxygenase enzymes that have catalytic activity towards aromatics, alkanes and MTBE. Confirming that strain VB-1 expresses such a novel monooxygenase is an area that requires further research.

Another area that requires further research is the pathway of MTBE oxidation by strain VB-1 after growth on aromatics. A pathway of MTBE oxidation to CO_2 has been proposed for alkane-oxidizing bacteria (Steffan *et al.*, 1997) which relies upon enzyme activities known to be associated with alkane oxidation. Our evidence presented here for strain VB-1 also suggests the initial steps in the oxidation of MTBE and TBA are the same in cells grown on alkanes and aromatics. However, the enzymes associated with the later steps of alkane and aromatic hydrocarbon oxidation are very different. It is therefore reasonable to expect that the pathway of MTBE oxidation after the monooxygenation of TBA may be very different in cells grown on aromatics and alkanes.

The ability of strain VB-1 to oxidize MTBE after growth on aromatics is interesting from the biochemical perspective and clearly raises some questions about the mechanism of MTBE degradation. However, perhaps the most significant feature of our observations with strain VB-1 is that it extends the substrates that are known to support MTBE cometabolism to include the two most significant aromatic compounds found in gasoline. Benzene is a relatively minor gasoline component but is of considerable significance because of its known carcinogenic properties. In contrast, toluene is considerably less toxic than benzene, but is found in very high concentrations in gasoline. Both toluene and benzene are known to degrade under aerobic and anaerobic conditions and bioremediation processes are often designed to maximize the rate of biodegradation of these compounds. Frequently the bioremediation of benzene and toluene and other members of the BTEX class of compounds is promoted by oxygenation. Our present results with strain VB-1 suggest that oxygenation of anaerobic environments containing toluene, benzene and MTBE may also promote MTBE cometabolism.

This research has been supported by a grant from the American Petroleum Institute to MRH. The views expressed here do no necessarily reflect the views of the American Petroleum Institute, and no official endorsement should be inferred.

REFERENCES

Hyman, M. R., P. Kwon, K. Williamson, and K. O'Reilly 1998 "Cometabolism of MTBE by Alkane-Utilizing Microorganism". In G.B. Wickramanayake and R. E. Hinchee (Eds.), *Natural Attenuation: Chlorinated and Recalcitrant Compounds*, pp. 321-326, Battelle Press, Columbus, OH.

Hyman, M. R., and K O'Reilly 1999 "Physiological and Enzymatic Features of MTBE-degrading Bacteria". In B. C. Alleman and A. Leeson (Eds.), *In Situ Bioremediation of Petroleum Hydrocarbons and Other Organic Compounds*, pp. 7-12, Battelle Press, Columbus, OH.

Hyman, M. R., C. Taylor, and K. O'Reilly 2000 " Cometabolic Degradation of MTBE by Iso-Alkane-Utilizing Bacteria from Gasoline-Impacted Soils". In G. B. Wickramanayake, A. R. Gavaskar, and B. C. Alleman and V. S. Magar (Eds.), *Bioremediation and Phytoremediation of Chlorinated and Recalcitrant Compounds*, pp 149-155, Battelle Press, Columbus, OH.

Steffan, R. J., K. McClay, S. Vainberg, C.W. Condee, and D. Zhang 1997 "Biodegradation of the Gasline Oxygenates Methyl *tert*-Butyl Ether, Ethyl *tert*-Butyl Ether and *tert*-Amyl Methyl Ether by Propane-Oxidizing Bacteria. *Appl. Environ. Microbiol.* 63 (11) 4216-4222.

SELECTION OF A DEFINED MIXED CULTURE FOR MTBE MINERALIZATION

Alan François[a], Pascal Piveteau[a], Françoise Fayolle[a], Rémy Marchal[a], Pierre Beguin[b] and **Frédéric Monot**[a]

[a] : Institut Français du Pétrole, Département de Microbiologie, 1-4, avenue Bois-Préau, 92852 Rueil-Malmaison Cedex, France.

[b] : Institut Pasteur, Unité de Microbiologie et Environnement, 25 rue du Docteur-Roux, 75015 Paris, France.

ABSTRACT: Mixed cultures of two defined bacterial strains have been used to study the cometabolic biodegradation of methyl *tert*-butyl ether (MTBE). The first strain, *Gordonia terrae* IFP 2007, was able to oxidize MTBE to *tert*-butyl alcohol (TBA) by cometabolism when grown on a carbon substrate such as ethanol or isopropanol. Another strain was necessary for the degradation of TBA produced to CO_2. For this purpose, two strains were isolated from activated sludge, *Burkholderia cepacia* IFP 2003 and *Mycobacterium sp.* IFP 2012. We compared the capacity of MTBE degradation of the two reconstituted mixed cultures, *G. terrae* IFP 2007/*B. cepacia* IFP 2003, and *G. terrae* IFP 2007/*Mycobacterium sp.* IFP 2012. In addition, two growth substrates were tested, ethanol and isopropanol. Both co-cultures were able to completely mineralize MTBE with high rates when using ethanol. Differences between co-cultures were observed, especially with isopropanol, since TBA disappeared in 120 h using *B. cepacia* IFP 2003 as a second strain and in 100 h using *Mycobacterium sp.* IFP 2012. However, these differences remained limited and both mixed cultures may be suitable for the implementation of a bioremediation process aimed at treating MTBE-contaminated plumes

INTRODUCTION

The presence of methyl *tert*-butyl ether (MTBE) in groundwater has become a major environmental concern. Recalcitrance of MTBE to biodegradation is due to the presence of both an ether bond and a tertiary carbon. Salanitro *et al.* (1994) obtained the first microbial consortium, BC-1, growing on MTBE as a sole carbon and energy source. Later, the cometabolic biodegradation of MTBE in the presence of short chain-alkanes was reported (Garnier *et al.*, 1999, Hardison *et al.*, 1997, Hyman *et al.*, 1998, Steffan *et al.*, 1997). Recently, two pure strains able to grow on MTBE as a sole carbon and energy source have been isolated, *Rubrivirex* sp. PM-1 (Deeb *et al.*, 2000) and *Hydrogenophaga flava* ENV735 (Steffan *et al.*, 2000).

At IFP, a pure bacterial strain, *Gordonia terrae* IFP 2007 growing on ethyl *tert*-butyl ether (ETBE), another fuel oxygenate, and producing *tert*-butyl alcohol (TBA), was isolated. In addition, this strain was able to degrade MTBE by cometabolism in presence of a growth substrate such as ethanol (Hernandez *et al.*, 2001). A strain degrading TBA to CO_2, *Burkholderia cepacia* IFP 2003, was also isolated (Piveteau *et al.*, 2001). Reconstituted mixed culture of both these strains was used to completely degrade ETBE (Fayolle *et al.*, 1999) and MTBE in presence of ethanol (Piveteau *et al.*, 2000). Another strain, *Mycobacterium sp.* IFP 2012, also was isolated from an aerobic activated sludge for its capacity to grow on TBA as a carbon and energy source.

Aiming at determining the best mixed culture, we studied the capacity of the two different bacterial associations, *G. terrae* IFP 2007/*B. cepacia* IFP 2003 and *G. terrae* IFP

2007/*Mycobacterium sp.* IFP 2012, to degrade MTBE in the presence of two growth substrates, ethanol or isopropanol (isopropyl alcohol).

MATERIALS AND METHODS

Microorganisms. *G. terrae* IFP 2007 was a constitutive variant of *G. terrae* IFP 2001 regarding growth on ETBE (Hernandez et al.,2001). *B. cepacia* IFP 2003 was isolated from an activated sludge for its ability to grow on TBA (Piveteau et al., 2001). *Mycobacterium sp.* IFP 2012 was also isolated from an activated sludge. The three strains were identified, deposited at the CNCM, Pasteur Institute, Paris, France, and stored in a glycerol solution (20%, w/w) at –80°C.

Growth conditions. Pure strains, *G. terrae* IFP 2007, *B. cepacia* IFP 2003 and *Mycobacterium sp.* IFP 2012, as well as the mixed cultures were cultivated on mineral medium (MM) as previously described (Piveteau et al., 2001). The solid medium used to check the purity of cultures was the Luria-Bertani (LB) medium supplemented with 20 $g.L^{-1}$ of pure agar. Typical orange colonies of *G. terrae* IFP 2007 were counted on LB plates. The specific determination of *B. cepacia* IFP 2003 and *Mycobacterium sp.* IFP 2012 counts was carried out on Petri dishes containing MM supplemented with TBA (1 $g.L^{-1}$) and 20 $g.L^{-1}$ of highly pure agar (Sigma).

Preparation of the different mixed cultures. Precultures of *G. terrae* IFP 2007 were prepared in 1 L of MM containing 1.5 $g.L^{-1}$ of ETBE. Precultures of *B. cepacia* IFP 2003 and *Mycobacterium sp.* IFP 2012 were carried out in 300 mL of MM medium containing 1.5 $g.L^{-1}$ TBA. Each culture was centrifuged at 10,000 g for 20 minutes and the cells were then washed and suspended in 10 mL of MM for seeding. Culture flasks containing 100 mL of MM medium were inoculated with 0.4 mL of *G. terrae* IFP 2007 suspension and 0.8 mL of *B. cepacia* IFP 2003 or *Mycobacterium sp.* IFP 2012. MTBE (70 $mg.L^{-1}$) and the growth substrate, *i.e.* ethanol or isopropanol (350 $mg.L^{-1}$), were added to the different cultures. Cultures were incubated in a rotary shaker at 30°C.

Analytical procedures. Culture samples were filtered (0.22 µm) and MTBE, TBA, ethanol or isopropanol concentrations were determined using a VARIAN 3300 gas chromatograph fitted with a flame ionization detector and equipped with a 0.32mm x 25m CP-Porabond Q capillary column (5 µm film thickness). The temperature of the column increased from 105°C to 200°C at 10°C.min^{-1}. The carrier gas was helium (1.6 $mL.min^{-1}$). The initial injector temperature was 110°C and then increased to 250°C at 180°C.min^{-1}. The temperature of the detector was 280°C.

Growth was followed by measuring the absorbance at 600 nm on a spectrophotometer (Shimadzu). The cell dry weight was determined from 50 to 100 mL of culture using a Mettler infra-red dryer (LP16) and weighing scale (PM100). Specific growth rates were determined from growth curves obtained by O.D.$_{600\ nm}$ measurements. Specific degradation rates are expressed per g cell dry weight.

Chemicals. ETBE, TBA, ethanol, isopropanol and acetone were purchased from Sigma.

RESULTS

MTBE degradation by *G. terrae* IFP 2007. The degradation activity of *G. terrae* IFP 2007 on MTBE was determined in presence of ethanol in sealed flasks (120 mL) containing 50 mL of

MM (Piveteau et al., 2000). The MTBE degradation rate was 32.3 mg.g^{-1}.h^{-1} and 110 mg.L^{-1} of MTBE were totally consumed in 25 hours at an ethanol/MTBE ratio of 4.1 (w/w).

TBA degradation by *B. cepacia* IFP 2003 and *Mycobacterium sp.* IFP 2012. Maximum specific growth rates (μ_{max}) and TBA degradation rates of both TBA-degrading strains are presented in Table 1. Although *Mycobacterium sp.* IFP 2012 grew more quickly than *B. cepacia* IFP 2003, the specific TBA degradation rates of both strains were quite similar.

TABLE 1. Growth rate and TBA degradation activity of *B. cepacia* IFP 2003 and *Mycobacterium sp.* IFP 2012.

Microorganism tested	μ_{max} in h^{-1}	TBA degradation rate in mg.g^{-1}.h^{-1}
B. cepacia IFP 2003	0.033	36 ± 8 (*)
Mycobacterium sp. IFP 2012	0.048	39 ± 2 (**)

(*) : from growing cells at an initial TBA concentration of 1 g.L^{-1}.
(**) : from resting cells (0.56 g.L^{-1} dry weight) at an initial TBA concentration of 200 mg.L^{-1}.
Experiments carried out in triplicate.

MTBE degradation by the mixed culture using ethanol as a growth substrate. The performances of the two mixed cultures, consisting of *G. terrae* IFP 2007 for MTBE degradation and *B. cepacia* IFP 2003 or *Mycobacterium sp.* IFP 2012 for TBA mineralization were compared (Figure 1).

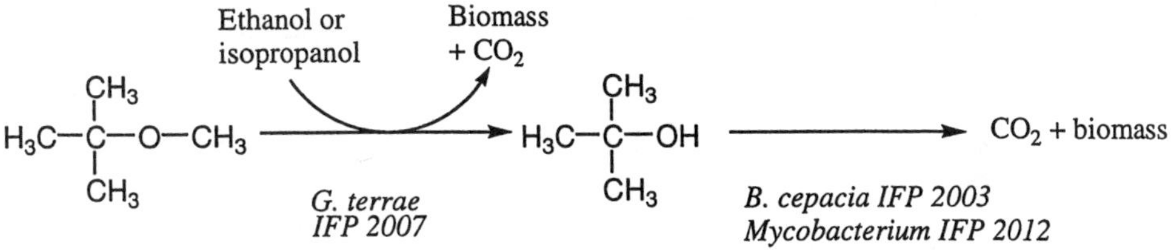

$$\text{MTBE} \xrightarrow[\text{G. terrae IFP 2007}]{\text{Ethanol or isopropanol} \rightarrow \text{Biomass} + CO_2} \text{TBA} \xrightarrow{\text{B. cepacia IFP 2003} \atop \text{Mycobacterium IFP 2012}} CO_2 + \text{biomass}$$

FIGURE 1. MTBE biodegradation by the reconstituted consortium *G. terrae* IFP 2007/*B. cepacia* IFP 2003 or *Mycobacterium sp.* IFP 2012

The initial cell number ratio was 1/20 for the culture *G. terrae*/*B. cepacia* and 1/25 for the culture *G. terrae*/*Mycobacterium sp.* The initial ethanol/MTBE ratio was around 4.5 (w/w). Time-courses of substrate and biomass concentration are shown on Figure 2.

Ethanol was totally consumed in 24 hours using both mixed cultures. Meanwhile, MTBE was degraded to TBA which transiently accumulated in the culture media. It was used by the TBA-degrading strain but more rapidly in the case of *Mycobacterium sp.* IFP 2012 than in the case of *B. cepacia* IFP 2003. TBA was nearly totally consumed after 48 hours (only 7 mg.L^{-1} of residual TBA at this time) in the first case and after 72 hours in the second one. The results showed that both mixed cultures were efficient for the MTBE degradation. However, the co-culture containing *Mycobacterium sp.* IFP 2012 was more efficient than the coculture containing

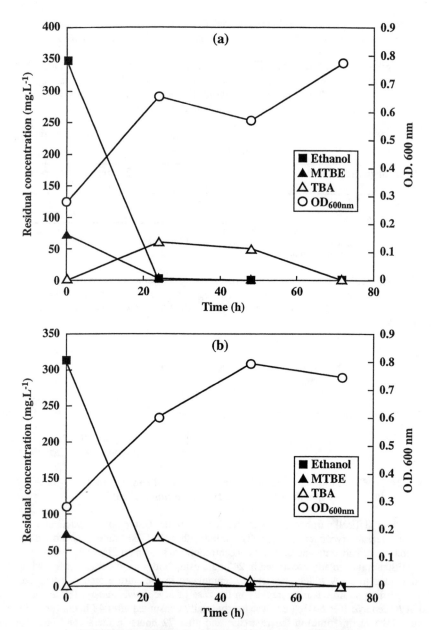

FIGURE 2 : MTBE degradation in presence of ethanol by the mixed cultures *G. terrae* IFP 2007/*B. cepacia* IFP 2003 (a) or *G. terrae* IFP 2007/ *Mycobacterium* sp. IFP 2012. (b)

B. cepacia IFP 2003.

MTBE degradation by the mixed culture using isopropanol as a growth substrate.

MTBE degradation by the two mixed cultures, G. terrae IFP 2007/B. cepacia IFP 2003 or Mycobacterium sp. IFP 2012 was compared in the presence of isopropanol. The initial cell number ratio was 1/20 for the culture G. terrae/B. cepacia and 1/17 for the culture G. terrae/Mycobacterium sp. The initial isopropanol/MTBE ratio was about 5.6 (w/w). Results are shown on Figure 3.

The use of isopropanol instead of ethanol as a growth substrate resulted in a longer time for complete MTBE degradation. No cell growth was detected during the first 24 hours while MTBE and isopropanol concentrations decreased. Therefore, after 24 hours, MTBE degradation yields were 17 and 29 % using the G. terrae/B. cepacia culture and the G. terrae/Mycobacterium sp.culture, respectively. Meanwhile, isopropanol degradation yields were 49 and 64 %, respectively. The decrease of isopropanol concentration was concomitant to the appearance of a new compound detected by gas chromatography (GC). This compound was identified as acetone by GC (co-elution in presence of pure acetone). The molar acetone/isopropanol ratio was 0.85 after 24 hours for each culture, showing that isopropanol was almost stoichiometrically oxidized to acetone. During the following 24 hours, oxidation of isopropanol to acetone was total. The acetone/isopropanol molar ratio at 48 h was 0.89 in the case of G. terrae/B. cepacia culture and 0.57 in the case of G. terrae/Mycobacterium sp. culture suggesting that this culture partially used acetone during this period. In the case of the G. terrae/B. cepacia culture, 82% of MTBE was degraded between 48h and 96h and this degradation was simultaneous to acetone consumption. In the case of the G. terrae/Mycobacterium sp. culture, MTBE was degraded more rapidly. TBA accumulation was limited using B. cepacia as the second strain. A transient TBA accumulation was observed using the G. terrae/Mycobacterium sp. culture. In both cases, the growth phase occurred during the acetone utilization phase showing that acetone was the actual growth substrate of G. terrae IFP 2007.

DISCUSSION

The accumulation of TBA during MTBE biodegradation by pure strains is frequently reported. The association of at least two strains is then required to completely and rapidly mineralize MTBE. The use of a defined mixed culture containing two selected strains would allow a good understanding and control of MTBE biodegradation processes. Furthermore, as MTBE is often degraded by cometabolism, the implementation of a bioremediation process also requires the presence of a carbon source different from MTBE for the growth of the MTBE-degrading strain. In the present case, G. terrae IFP 2007 can degrade MTBE to TBA in the presence of ethanol or isopropanol. In addition, two pure strains utilizing TBA as a carbon and energy source have been isolated. The selection of the most suitable system needed defining both the best carbon source for G. terrae and the best TBA-degrading strain.

The results showed that higher MTBE degradation rates were obtained using ethanol. Nevertheless, considering the implementation of an in-situ bioremediation process, isopropanol is probably a more selective substrate than ethanol regarding its possible utilization as growth substrate by indigenous soil microorganisms. The results of the present study showed that G. terrae IFP 2007 first converted isopropanol to acetone which was further used for its growth.

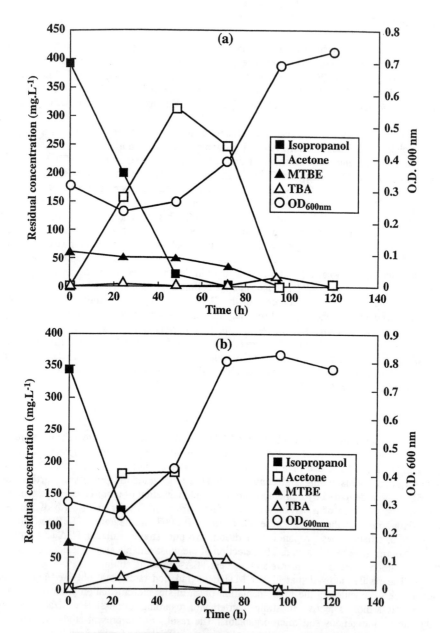

FIGURE 3 : MTBE degradation in presence of isopropanol by the mixed cultures *G. terrae* IFP 2007/*B. cepacia* IFP 2003 (a) or *G. terrae* IFP 2007/*Mycobacterium* sp. IFP 2012. (b)

This result also suggests that it would be interesting to consider acetone as another possible carbon and energy source for *G. terrae* IFP 2007. Both *Mycobacterium sp.* IFP 2012 and *B. cepacia* IFP 2003 exhibit high TBA degradation rates. Concerning the choice of the most suitable co-culture, *B. cepacia* IFP 2003 and *Mycobacterium sp.* IFP 2012 are convenient since they rapidly degraded the TBA formed by *G. terrae* IFP 2007. The rate of TBA utilization by these strains was lower than the TBA production rate of *G. terrae* and TBA transiently accumulated in the medium. However, *Mycobacterium sp.* IFP 2012 strain was preferable to *B. cepacia* IFP 2003 to achieve high degradation rates, especially in the presence of isopropanol. It is possible that the isopropanol-oxidation product, acetone, was used by both *G. terrae* IFP 2007 and *Mycobacterium sp.* IFP 2012. This latter point would deserve further investigation. In conclusion, this study shows that these mixed cultures of two selected strains are efficient with respect to MTBE degradation and that their use in bioremediation processes can be considered at a larger-scale.

REFERENCES

Deeb, R.A., S. Nishino, J. Spain, H.-Y. Hu, K. Scow and L. Alvarez-Cohen. 2000. MTBE and Benzene Biodegradation by a Bacterial Isolate via Two Independent Monooxygenase-Initiated Pathways. Preprints of Extended Abstracts, ACS Natl. Meet., Am. Chem. Soc., Div. Environ. Chem 40(1), pp. 280-282.

Fayolle, F, F. Le Roux, G. Hernandez and J.-P. Vandecasteele. 1999. Mineralization of Ethyl *t*-Butyl Ether by Defined Mixed Bacterial Cultures. *In* : B.C. Aleman and A. Leeson (Ed(s).), In Situ Bioremediation of Petroleum Hydrocarbon and Other Organic Compounds. Battelle Press, Columbus, OH, pp.25-30.

Garnier, P., R. Auria, M. Magana, and S. Revah. 1999. Cometabolic Biodegradation of Methyl *t*-Butyl Ether by a Soil Consortium. *In* : B.C. Aleman and A. Leeson (Ed(s).), In Situ Bioremediation of Petroleum Hydrocarbon and Other Organic Compounds. Battelle Press, Columbus, OH, pp.31-35.

Hanson, J.R., C.R. Ackerman, and K.M. Scow. 1999. Biodegradation of Methyl *tert*-Butyl Ether by a Bacterial Pure Culture. *Appl. Environ. Microbiol.*, 65, 4788-4792.

Hardison, L.K., S.S. Curry, L.M. Ciuffetti and M.L. Hyman. 1997. "Metabolism of Diethyl Ether and Cometabolism of MTBE by a Filamentous Fungus, a *Graphium sp.*". *Appl. Environ. Microbiol.*, 63, 3059-3067.

Hernandez, G., F. Fayolle and J.-P. Vandecasteele. 2001. "Biodegradation of Ethyl *t*-Butyl Ether (ETBE), Methyl *t*-Butyl Ether (MTBE) ant *t*-Amyl Metyl Ether (TAME) by *Gordona terrae*." *Appl.Microbiol.Biotechnol.*, 55, 117-121.

Hyman, M., P. Kwon, K. Williamson and K. O'Reilly. 1998. Cometabolism of MTBE by Alkane-Utlizing Microorganisms. *In* : G.B. Wickramanayake and R.E. Hinchee (Ed(s).), Natural Attenuation of Chlorinated and Recalcitrant Compouds. Battelle Press, Columbus, OH, pp. 321-326.

Piveteau P., F. Fayolle, Y. Le Penru and F. Monot. 2000. Biodegradation of MTBE by Cometabolism in Laboratory-Scale Fermentations. *In* : G.B. Wickramanayake, A.R. Gavaskar, B. Alleman and V.S. Magar (Ed(s).), Bioremediation and Phytoremediation of Chlorinated and Recalcitrant Compounds. Battelle Press, Columbus, OH, pp. 141-148.

Piveteau P., F. Fayolle, J.-P. Vandecasteele and F. Monot. 2001. "Biodegradation of *t*- Butyl Alcohol and Related Xenobiotics by a Methylotrophic Bacterial Isolate. " *Appl.Microbiol.Biotechnol.*, 55, 369-373.

Salanitro, J.P., L.A. Diaz, M.P. Williams and H.L. Wisniewski. 1994. "Isolation of a Bacterial Culture that Degrades Methyl *t*-Butyl Ether". *Appl. Environ. Microbiol.*, 60, 2593-2596.

Steffan, R.J., K. Mac Clay, S. Vainberg, C.W. Condee and D. Zhang. 1997. "Biodegradation of the Gasoline Oxygenates MTBE, ETBE and TAME by Propane-Oxidizing Bacteria. *Appl. Environ. Microbiol.*, 63, 4216-4222.

Steffan R., S. Vainberg, C. Condee, K. Mc Clay and P. Hatzinger. 2000. Biotreatment of MTBE with a New Bacterial Isolate. *In* : G.B. Wickramanayake, A.R. Gavaskar, B. Alleman and V.S. Magar (Ed(s).), Bioremediation and Phytoremediation of Chlorinated and Recalcitrant Compounds. Battelle Press, Columbus, OH, pp. 165-173.

BIODEGRADATION OF MTBE AND OTHER GASOLINE OXYGENATES BY BUTANE-UTILIZING MICROORGANISMS

Soon W. Chang, Kyonggi University, Korea
Seung S. Baek, Kyonggi University, Korea
Si J. Lee, Kyonggi University, Korea

ABSTRACT: In this study, we present oxidation of MTBE (methyl *tert*-butyl ether) and other gasoline oxygenates by butane-grown microorganisms, ENV425 and a mixed culture. Our results indicate that butane-grown bacteria ENV425 and the mixed culture can cometabolically degrade MTBE and other gasoline oxygenates. The degradation of MTBE and other gasoline oxygenates was completely inhibited by acetylene and strongly influenced by the addition of *n*-butane. Two major products of MTBE degradation, *tert*-butyl formate (TBF) and *tert*-butyl alcohol (TBA), were detected in the resting cell experiment while TBA was major product in the simultaneous degradation of *n*-butane and MTBE. The kinetics of products formation suggest that TBF production rapidly proceeds to TBA. Our results suggest that MTBE oxidation is initiated by butane-monooxygenase reactions, which lead to scission of the ether bond in this compound. Our findings also suggest a potential role for gaseous *n*-alkane-oxidizing bacteria in the remediation of MTBE contamination.

INTRODUCTION

Methyl *tert*-butyl ether (MTBE) is an octane enhancer commonly used in reformulated gasoline. The use of MTBE is still increasing in Korea, whereas it is reducing in the United State due to its potential health risk.

Salanitro et al. (1994) observed the first microbial consortium grown on MTBE as a sole carbon and energy source. Other researchers have focused on cometabolic processes in which MTBE is degraded by microorganisms that grew on other substrates. Hyman et al. (1999, 2000) have previously demonstrated that several organisms can degrade MTBE after growth on a variety of low molecular weight alkanes, including straight chain (*e.g.* propane, *n*-butane and *n*-pentane).

In this study, we examined the ability of butane-oxidizing bacteria isolated from aquifer materials in Korea, to metabolize gasoline oxygenates, including MTBE, ETBE, TAME, and DIPE. Because *n*-alkane-oxidizers are widespread in nature as described by several researchers, they may play an important role for the degradation of MTBE and other oxygenates in soils and groundwater, and they may provide the basis for a potential biological treatment process.

MATERIALS AND METHODS

Materials. ENV425 (ATCC55798) was obtained from the Environ. Inc. (U.S.A.).

MTBE (99%), ETBE (97%), TAME (94%) and DIPE (99+%) were obtained from ACROS ORGANICS (New Jersey, U.S.A). TBA (99%) and TBF (99%) were obtained from Sigma Chemical Co. (St. Louis, MO, U.S.A.) and Aldrich Chemical Co. (Milwaukee, U.S.A.), respectively. All gases were of the highest purity available, and all other chemicals were at least of reagent grade

Growth Conditions. ENV425 was grown as previously described (Steffan, 1997). And the mixed culture was derived using soil from a gasoline-contaminated site in Korea. An aseptically collected soil sample (3g) was placed in a batch reactor containing n-butane (5 mL) as the sole carbon and energy source and mineral salts medium with a media described previously (Steffan, 1997). Once bacterial growth was observed in the aqueous phase after 3 weeks, a liquid sample (100 µL) was used to inoculate a second batch system. This procedure was repeated several times and it was used for batch experiments.

Degradation Experiments. Microcosm studies were conducted in sterilized serum bottles (120 mL) with Teflon-coated silicone septa and aluminum crimp caps. MTBE and other oxygenates were injected from a stock solution prepared in water with a 100- and 1000-µL syringes. The gaseous hydrocarbon substrate was added to the bottles as an overpressure by using syringes fitted with sterile filters (0.25 µm pore size). Liquid suspension cultures were incubated for 7~8 days at 25°C in an orbital shaker (150 rpm). The cells were grown on n-butane to an OD_{550} of 0.4 and then collected by centrifugation, washed with BSM, and resuspended in sodium buffer phosphate buffer (50 mM, pH 7.0) for the experiment. These cell suspensions were stored on ice and were used in the experiments.

The degradation of MTBE and other oxygenates was investigated in 120 mL serum bottles maintained at 25°C in a shaking chamber (150 rpm). The reactions were initiated in serum bottles (120 mL) stoppered with butyl rubber septa and aluminum crimp seals. The reaction bottles contained BSM (50 mL) and MTBE and other oxygenates (2 µmols). The reactions were initiated by the addition of cells (7.0 to 10.0 mg dry weight) to the bottles. The gas samples (100 µL) were removed from the reaction bottles and were directly injected into gas chromatography. Also liquid samples (500 µL) were removed and then were extracted CS_2 (Carbon Sulfide) to determine TBA concentrations.

RESULTS AND DISCUSSIONS

The growth yields for the ENV425 and mixed cultures on n-butane were determined by using substrate-limited batch cultures in which the biomass was determined after the complete consumption of added n-butane had been confirmed by gas chromatography. These observed yields were 56.9 and 54.1 (g of biomass/mol of n-butane consumed) in the ENV425 and mixed culture, respectively and this reaction was inhibited by acetylene. Neither culture was grown on MTBE as the sole source of carbon and energy. The maximal rate of n-butane oxidation in this experiment was 62.3 nmol oxidized/h/mg (dry weight).

Figure 1 shows MTBE degradation and TBA or TBF production in ENV 425 and the mixed culture and both cultures were unable to be grown on MTBE as a sole source of carbon and energy. However, n-butane-grown cultures were capable of degrading MTBE by cometabolic reaction, and also consumption of MTBE by n-butane-grown cultures was fully inhibited by the acetylene (data not shown), which inactivates n-butane-oxidizing activity. The effects indicate that the consumption of MTBE was due to butane-monooxygenase activities rather than to nonspecific or abiotic processes.

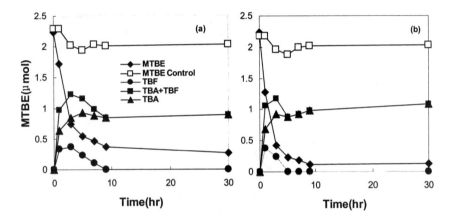

FIGURE 1. Time courses for TBA and TBF production during MTBE degradation without n-butane by n-butane-grown ENV425 (a) and mixed culture (b). MTBE (2 µmol) was incubated with ENV425 and mixed culture (9.4mg [dry weight]) in a sealed glass amber serum bottle (120 mL). At the indicated times, sample (100 µL) of the gas phase were removed and immediately analyzed by gas chromatography. The symbol (□) is indicated the MTBE degradation in the presence of acetylene (1.0% (vol/vol); gas phase).

We also examined the possible accumulation of MTBE oxidation products in experiments with ENV425 and mixed culture grown on n-butane. In these experiments, we consistently detected only one product, which was identified as TBA by co-elution with the authentic compounds during gas chromatography. The kinetics of TBA production during MTBE consumption were investigated, and the results indicated that TBA and TBF production could not accounted for all of the MTBE consumption by ENV425 and mixed cultures.

In the resting cell experiments, we consistently detected two products, which were identified as TBF and TBA as shown in the Fig. 1. We also determined transformation capacity of washed butane-grown cells. The effects of both cell density and MTBE concentration on transformation capacity were studied. Cells were incubated with MTBE for 15 h, after which butane-oxidizing activity was detected. Table 1 shows the amounts of MTBE oxidized and transformation capacity by the added biomass and MTBE concentration.

TABLE 1. Transformation capacity on the MTBE degradation as the functions of cell density and initial MTBE concentration.

Culture	Amount of Biomass (mg/mL)	Amount of MTBE degraded at the following Concentration of MTBE					
		2µmol		8µmol		16µmol	
		MTBE (µmol)	T_c	MTBE (µmol)	T_c	MTBE (µmol)	T_c
ENV425	0.14	1.09±0.03	0.15±0.003	3.41±0.15	0.48±0.020	4.78±0.06	0.67±0.010
	0.28	1.07±0.21	0.08±0.015	2.57±0.04	0.18±0.003	5.55±0.62	0.39±0.042
	1.42	1.18±0.09	0.06±0.004	4.05±0.18	0.19±0.004	8.35±0.43	0.39±0.018
Mixed Culture	0.14	0.84±0.22	0.12±0.031	2.56±0.12	0.36±0.016	2.91±0.14	0.41±0.014
	0.28	0.79±0.04	0.06±0.003	2.79±0.08	0.20±0.005	4.39±0.32	0.31±0.021
	1.42	1.19±0.17	0.06±0.008	4.11±0.07	0.19±0.003	6.88±0.75	0.32±0.032

The results show that the transformation capacity decreased with cell density. However, increasing MTBE concentration increased the transformation capacity. The highest transformation capacity was observed with the higher MTBE concentration and lower cell density.

To test the ability of butane-utilizing cultures to degrade other gasoline oxygenates, cells were grown as described in materials and methods. Both cultures tested degraded MTBE, ETBE, TAME, and DIPE. As shown in Table 2, butane-utilizing bacteria oxidized all the other oxygenates tested. But extent of degradation by ENV425 and the mixed culture was not consistent.

TABLE 2. Degradation of MTBE and other oxygenates by ENV425 and mixed culture grown on *n*-butane after 10 hr incubation

Culture	Compounds	*n*-butane degradation µmol (%)	Gasoline Additives degradation µmol (%)
ENV425	MTBE	45.6 (83.3)	1.3 (89.6)
	ETBE	49.4 (81.8)	1.2 (60.8)
	TAME	48.8 (86.4)	1.2 (74.9)
	DIPE	46.3 (83.2)	1.2 (54.0)
Mixed Culture	MTBE	39.6 (64.9)	0.8 (47.4)
	ETBE	41.6 (78.1)	1.2 (75.3)
	TAME	46.2 (84.1)	1.4 (97.8)
	DIPE	47.1 (87.9)	1.8 (91.0)

ACKNOWLEDGEMENT

The author would like to thank Robert Steffan (Envirogen Inc., USA) for providing strain ENV425 for our experiment. We would also like to thank Korea Research Foundation (KRF) and Kyonggi University for their financial support for this study.

REFERENCES

Hyman, M.R., P. Kwon, K. Williamson, and K. O'Reilly. 1998. "Cometabolism of MTBE by Alkane-Utilizing Microorganisms", *In*. Wickramanayake, G.B. and R.E. Hinchee (ed.), *Natural Attenuation: Chlorinated and Recalcitrant Compounds*, pp.321-326, Battelle Press, Columbus, OH.

Hyman, M., C. Taylor, and K. O'Reilly. 2000. "Cometabolic Degradation of MTBE by Iso-Alkane-Utilizing Bacteria from Gasoline-Impacted Soils", *In*. G.B. Wickramanayake, A.R.Gavaskar, B.C. Alleman, and V.C. Magar (ed.), *Bioremediation and Phytoremediation of Chlorinated and Recalcitrant Compounds*. Pp.149-155, Battelle Press, Columbus, OH.

Salanitro, J.P., L.A. Diaz, M.P. Williams, and H.L. Wisniewski. 1994. "Isolation of Bacterial Culture that Degrades Methyl *t*-Butyl Ether", *Applied and Environmental Microbiology*, 60:2593-2596.

Steffan, R.J., K. McClay, S. Vainberg, C.W. Condee, and D. Zhang. 1997. "Biodegradation of the Gasoline Oxygenates Methyl *tert*-Butyl Ether, Ethyl *tert*-Butyl Ether, and *tert*-Amyl Methyl Ether by Propane-Oxidizing Bacteria", *Applied and Environmental Microbiology*, 63(11):4216-4222.

IMPACT OF ETHANOL ON BENZENE MIGRATION

Hai Shen (Dynamac Corporation, Ada, Oklahoma)
Susan C. Mravik, John T. Wilson and Guy W. Sewell (U.S. EPA, Ada, Oklahoma)

ABSTRACT: The use of ethanol as a fuel oxygenate is expected to expand rapidly due to MTBE contamination in ground water. It is essential to understand the potential impact of ethanol-blended gasoline releases to ground water quality. The presence of ethanol is likely to impact the length and persistence of a benzene plume. Calculations using available field and laboratory data show that ethanol biodegradation rate constants varied from 0.045 to 34 mg L^{-1} d^{-1}. Depending on the ethanol biodegradation rate constants, it is estimated that a benzene plume in the subsurface could expand from unapparent to a very large extent in the presence of ethanol compared to cases without ethanol. The results may serve as both upper and lower bounds in evaluating the impact of ethanol on benzene migration.

INTRODUCTION

It is well known that biodegradation by native subsurface bacteria is the primary mechanism for natural attenuation of fuel hydrocarbons, including benzene, toluene, ethylbenzene, and the xylenes (BTEX), in aquifers. However, natural biodegradation of fuel hydrocarbons, particularly BTEX, which includes the major contaminants of concern, may change significantly when the released gasoline contains ethanol as a fuel oxygenate. The presence of ethanol is most likely to impact the length and persistence of the BTEX plume due to its potential impact on natural biodegradation in the subsurface. Ethanol, as a preferentially biodegradable compound, could potentially delay the biodegradation of BTEX compounds, which would increase the length of BTEX plume.

It is well documented that biodegradation of ethanol can easily occur under both aerobic and anaerobic conditions, including denitrifying, iron-reducing, sulfate-reducing and methanogenic conditions (Suflita and Mormile, 1993; Hunt et al., 1997). It is expected that large quantities of ethanol released from spills of ethanol-blended gasoline will easily deplete available oxygen in ground water. Thus, evaluation of the effects of ethanol on natural biodegradation of BTEX should focus on anaerobic biodegradation, dependent on electron acceptors such as nitrate, sulfate, and iron. To date, only limited laboratory information regarding the effect of ethanol on the biodegradation of BTEX is available, and as far as we are aware, there is only one field study. Three recent modeling studies predicted a potential extension of benzene plume lengths from 20% to 180% (Malcolm Pirnie, Inc., 1998; Ulrich, 1999; McNab et al., 1999). It is not surprising that these three independent studies generated similar results because they obtained ethanol biodegradation rates based on the same laboratory data.

The ethanol biodegradation rate is a critical parameter in determining the impact of ethanol on BTEX biodegradation in the subsurface. Laboratory and field data were employed to estimate an ethanol biodegradation rate constant which is used, in turn, to estimate the potential impact of ethanol on benzene biodegradation in the subsurface. It is believed that the results provide a complementary approach to that used in earlier studies for estimating the impact of ethanol on natural biodegradation of BTEX. For simplicity, benzene is used as a representative compound here because it is the most resistant BTEX compound to anaerobic biodegradation and usually presents the greatest risk to human health.

INITIAL BENZENE CONCENTRATIONS IN GROUND WATER

Data collected by EPA (Ravi, et al., 2000) were used to extract a high, medium and low estimate of the initial concentration of benzene in ground water impacted by gasoline spills. The concentration of benzene in monitoring wells at 74 gasoline service stations, primarily from the mid-Atlantic region of the United States, was analyzed. The distribution in maximum benzene concentrations detected at each of the 74 gasoline service station sites was used to estimate the initial concentration of benzene at the source of the plume. Maximum benzene concentrations at the 74 gasoline service stations ranged from 0.5 to 32,700 µg/L. The median benzene concentration was 360 µg/L, and the lower (25%) and upper (75%) quartile benzene concentrations were 98 µg/L and 4390 µg/L, respectively. These three values will be used as benzene source concentrations to evaluate the impact of ethanol on natural anaerobic biodegradation of benzene.

ESTIMATE OF BENZENE BIODEGRADATION RATE CONSTANT

In a recent review article of natural biodegradation of organic compounds in ground water, the mean of the first order rate constants for in situ anaerobic biodegradation of benzene was 0.003 per day with an upper 90th percentile value of the rate of 0.009 per day (Suarez and Rifai, 1999). Both values were used in the analysis to predict benzene biodegradation in the subsurface after ethanol had been depleted.

The study that was done in Florianopolis, SC, Brazil monitored a plume of BTEX and ethanol from a release of gasoline with 24% ethanol. They found that benzene did naturally attenuate in the presence of ethanol, but at a much slower rate. The rate reported from the study in Brazil was 0.00047 per day (Conseuil et al., 2000), which is only 15% of the mean of the rates summarized in the review article.

ESTIMATE OF ETHANOL BIODEGRADATION RATE CONSTANT

The ethanol biodegradation rate constant is a critical parameter for predicting the fate and transport of benzene in ground water. It is believed that the presence of ethanol will inhibit microbial consumption of benzene. The lower the rate of ethanol biodegradation, the longer the lag time before benzene biodegradation begins. A zero-order reaction for ethanol biodegradation in the subsurface will be assumed because high concentrations of ethanol are expected at the spill site. Easily degradable organic compounds at high concentrations have been observed to generally follow a zero-order reaction under anaerobic conditions. Data from field studies and laboratory microcosm studies of ethanol biodegradation were used for direct calculation of the zero order rate constant. Because behavior of ethanol and benzene will be extrapolated over longer periods of time than the time of incubation of the laboratory studies or the field study, one of the rate estimates was based on the total quantity of organic materials that had been metabolized at a gasoline spill site that was at least twenty years old.

Ethanol-Enhanced Remediation at Sage Site, Jacksonville, FL (Sage site). This site is contaminated with chlorinated solvents. The solvents were partially removed by flushing the site with a solution of 90 percent ethanol in water. At the completion of the flushing project, a solution of approximately five percent (by weight) ethanol was left in the subsurface along with residual chlorinated solvents that were not removed by the flushing project. The field experiment was originally designed to study the enhanced biodegradation of residual PCE using residual ethanol as an electron donor (Mravik et al., 2000). It also provides a good opportunity to gain insight into ethanol biodegradation in the subsurface environment. Assuming that the ethanol biodegradation at the Sage site followed a zero-order reaction gives an average ethanol biodegradation rate constant of 11 mg per liter per day. A conservatively lower estimate of the ethanol biodegradation rate constant is the lower 95% confidence interval which is 2.8 mg per liter per day.

Fuel-Contaminated Aquifer in Oklahoma City, OK (Oklahoma City site). The amount of Fe(II) and reduced S minerals that have accumulated at an old fuel spill site can be used to calculate the quantity of organic compounds that have been degraded by native sulfate-reducing and iron-reducing microorganisms during the entire history of the spill (Kennedy et al., 1998). In terms of a microbial process that depends on Fe(III) minerals as an electron acceptor, it is most likely that the biodegradation process is limited by availability of the electron acceptor. If the electron acceptor is assumed as the rate-limiting factor for natural biodegradation of fuel hydrocarbons during the past 15 years, the measured Fe(II) mineral generation rate as a result of natural biodegradation of fuel hydrocarbons at the site can be converted to the rate of natural biodegradation of ethanol according to stoichiometric relationship between Fe(II) and ethanol. Based on the accumulation of Fe(II) minerals, and by assuming the aquifer has a solid material density at 1.8 kg/L and porosity of 20%, the natural biodegradation rate constant for ethanol at this site is estimated to be equivalent to 0.045 mg per liter per day.

Experimental Release of Gasohol in Brazil (Brazil site). The only other data available to us from a field study with ethanol is the study conducted at Florianopolis, SC, Brazil (Conseuil et al., 2000). The effective first order decay constant for ethanol, corrected for dilution by attenuation of bromide, was 0.00115 per day. At the time of sampling the maximum concentration of ethanol remaining was 2503 mg/L. This rate multiplied by the maximum concentration yields a maximum instantaneous rate of 2.9 mg per liter per day. This is in agreement with the conservative lower estimate at the Sage site of 2.8 mg per liter per day. The authors of the study in Brazil chose to extract a first order rate constant for ethanol instead of a zero order rate constant. They also reported a slow but not negligible rate of benzene degradation in the presence of ethanol. As a result, the data from this site were evaluated as if there were no lag induced by the presence of ethanol, but using the lower rate constant for benzene decay.

Microcosm Studies. The first microcosm study is natural biodegradation of ethanol in aquifer slurries (Suflita and Mormile, 1993). The microcosm study was conducted using sediment and ground water collected from a methanogenic portion of a shallow anoxic aquifer in Norman, Oklahoma that was polluted by municipal landfill leachate. The estimate of the zero-order rate constant for ethanol degradation from this study was 34 mg per liter per day.

The second microcosm study is natural biodegradation of ethanol in subsurface sediment (Corseuil et al., 1998). The microcosm study under denitrifying and iron-reducing conditions was conducted with sandy soil from the Pentacrest area of the University of Iowa, Iowa City, Iowa. The estimate of the zero-order rate constant for ethanol degradation was 50 mg per liter per day under denitrifying conditions and 12.5 mg per liter per day under iron reducing conditions. The microcosm study under sulfate reducing conditions was conducted with sediments from a pond near Iowa City, Iowa. The estimate of the zero-order rate constant for ethanol degradation was 10 mg per liter per day under sulfate reducing conditions.

The rate of ethanol biodegradation estimated from different treatments in the two microcosm studies was 50, 34, 12.5 and 10 mg per liter per day. The value of 34 mg per liter per day was used as the median rate ethanol biodegradation rate constant for the microcosm studies.

EFFECT OF ETHANOL ON BENZENE MIGRATION

The conceptual model evaluated in this study was that ethanol is preferentially biodegraded, as compared with benzene, and that the biodegradation of benzene only begins after the ethanol is depleted. This preferential biodegradation of ethanol results in a lag time during which no biodegradation of benzene takes place. During the lag, a benzene plume is expected to expand without significant attenuation. Once the ethanol is depleted, the rate of benzene biodegradation is assumed to be constant and independent of whether or not ethanol was originally present. In the analysis, it was assumed that the length of a plume of benzene in the presence and absence of ethanol will be directly proportional to the time required for natural biodegradation to remove benzene to the MCL (5 µg/L). The assumption that no biodegradation of benzene takes place until ethanol is completely degraded is conservative according to the field observation of a slow benzene biodegradation in the presence of ethanol at the Brazil site.

In order to evaluate the effects of ethanol on a benzene plume in the subsurface, several assumptions were made in the study: 1) ethanol is present at a 10% volume in the ethanol-blended gasoline which will result in an ethanol concentration of 4,000 mg/L in ground water if released to the subsurface; 2) the concentration of ethanol in gasoline is not sufficient to cause a significant amount of cosolvent solubilization of benzene; 3) ethanol is preferentially biodegraded before benzene; 4) persistence of benzene in the environment when ethanol is present will result in enhanced movement of benzene; 5) benzene and ethanol are degraded following the first-order and zero-order reactions, respectively; 6) dispersion and sorption are ignored to simplify the evaluation.

Based on the kinetic parameter values obtained in the previous sections, the time that is required to reach a benzene concentration of 5 µg/L in ground water was calculated under various conditions (Table 1). In the absence of ethanol, the concentration of benzene was estimated to reach 5 ug/L after 0.9 to 6.2 years. When ethanol is present, and using the assumptions noted earlier, the time needed to achieve the benzene remedial goal varied from 1.2 to 10 years under more probable conditions. The worst case scenario resulted in times of up to 250 years.

TABLE 1. The effect of ethanol (4000 mg/l) on benzene migration.

Site Conditions	Lag Year	Years for Benzene to Reach 5 µg/L					
		$k_b = 0.003$ d^{-1} (mean)			$k_b = 0.009$ d^{-1} (90th)		
		C_0=98	360	4390	C_0=98	360	4390
No Ethanol	0	2.7	3.9	6.2	0.9	1.3	2.1
Sage Site (average), k_e=11 mg L^{-1} d^{-1}	1.0	3.7	4.9	7.2	1.9	2.3	3.1
Sage Site (the 95th), k_e=2.8 mg L^{-1} d^{-1}	3.9	6.6	7.8	10	4.8	5.2	6.0
Oklahoma City Site, k_e=0.045 mg L^{-1} d^{-1}	244	246	247	250	244	245	246
Microcosm Studies, k_e=34 mg L^{-1} d^{-1}	0.3	3.0	4.2	6.5	1.2	1.6	2.4
Brazil Site, k_e=2.9 mg L^{-1} d^{-1}	0.0	18	25	40	(values were calculated based on k_b=0.00047 d^{-1})		

C_0: initial benzene concentration (lower quartile=98 µg/L; median=360 µg/L; upper quartile=4390 µg/L); k_e: ethanol biodegradation rate constant; k_b: benzene anaerobic degradation rate constant.

If the ethanol biodegradation rate constant follows the average value (k_e=11 mg L^{-1} d^{-1}) estimated directly from the field ethanol concentrations observed at the Sage site, the time needed to achieve the benzene remedial goal by natural biodegradation will be from 1.9 to 7.2 years. The results suggest that a benzene plume in the aquifer will extend 16% further in the case of plumes with the highest concentration of benzene and a lower benzene degradation rate, and 110% further in the case of plumes with the lowest concentration of benzene and a higher benzene degradation rate. The results are very close to the previous evaluation of 25 to 180% expansion estimated by other investigators (Malcolm Pirnie, Inc., 1998; Ulrich, 1999; McNab et al., 1999), although the ethanol biodegradation rate constant was estimated through different approaches including laboratory and field studies for our study as compared with theirs. If the conservative lower value for the ethanol biodegradation rate constant is used (k_e is at least 2.8 mg L^{-1} d^{-1}, at 95% confidence), the time needed to achieve the benzene remedial goal by natural biodegradation will be from 4.8 to 10 years. The results also indicate that a benzene plume in the presence of ethanol will expand from 63% in the case of plumes with the highest concentration of benzene and a lower degradation rate, to 430% in the case of plumes with the lowest concentration of benzene and a higher degradation rate, compared to the scenario without ethanol.

If the ethanol biodegradation rate constant follows the value of 34 mg L^{-1} d^{-1} that was estimated from the two microcosm studies, the time needed to achieve the benzene remedial goal by natural biodegradation will vary from 1.2 to 6.5 years. This suggests that in the presence of ethanol, the benzene plumes will be 5% longer in the plumes with the highest concentration of benzene and a lower degradation rate, and 35% longer in the plumes with the lowest concentration of benzene and a higher degradation rate. Given the density of monitoring wells that are used at most sites, this increase would be barely detectable. The estimated values of benzene plume expansion are in the lower range predicted by other investigators (Malcolm Pirnie, Inc., 1998; Ulrich, 1999; McNab et al., 1999).

If the ethanol biodegradation rate constant follows the value (k_e=0.045 mg L^{-1} d^{-1}) estimated from the Oklahoma City site based on the assumption that Fe(III) reduction is the only bioremediation process taking place, the time needed to achieve the benzene remedial goal by natural biodegradation would be from 244 to 250 years. This long lag time would lead to a very large expansion of the benzene plume in the presence of ethanol compared to the cases without ethanol. This estimate of time required to meet the benzene MCL should be considered as the worst-case scenario.

This is because that the rate estimated from the Oklahoma City site, may be a very conservative lower limit on the rate of ethanol degradation for most sites where fuel spills are likely to occur. At many fuel spills, the contribution of Fe(III) reduction to biodegradation at fuel spill sites is small compared with that of other biodegradation processes. A recent study of 38 sites that had been contaminated with petroleum hydrocarbons indicates that the contribution of the Fe(III) reduction mechanism to BTEX removal is relatively small (2%) compared to the contributions by sulfate reduction (70%), methanogenesis (16%), denitrification (9%) and aerobic respiration (3%) (Wiedemeier et al., 1999). If Fe(III) reduction processes at the Oklahoma City site were assumed to account for only 2% of the total organic material degraded, the adjusted ethanol biodegradation rate constants at the sites becomes 2.3 mg per liter per day, which are more consistent with the rates of biodegradation of ethanol from microcosm studies and pilot scale field studies.

The small values for the rate constants were derived on the assumption that iron reducing processes are predominantly responsible for biodegradation and probably represent a worst case scenario of what could happen if ethanol-blended fuel is released to the subsurface. This scenario might occur if a large quantity of ethanol-blended fuel was released to a subsurface geological environment with very low permeability or limited supplies of sulfate and nitrate. It should also be noted that among the 38 sites reported in the recent study, Fe(III) reduction was not the dominant electron accepting process at any site. Although the 38 sites were not selected as a comprehensive survey of fuel contaminated sites in the United States, they did include sites which are representative of many areas of the United States. While a very low rate of ethanol biodegradation could occur, it seems that it would likely do so infrequently.

Uncertainty also results from the assumption that benzene biodegradation is altogether prevented by the presence of ethanol. The significance of this assumption can be evaluated by comparing the average time required to attenuate benzene in the absence of ethanol to the time required at a much slower rate of benzene degradation measured in the presence of ethanol at the Florianipolis, Brazil site. At the highest benzene concentration, degradation was projected to require 40 years compared to 6.2 years in the absence of ethanol. In the presence of ethanol, the benzene plumes would be 645% longer.

It is also important to emphasize that uncertainties remain. The original experiment at the Sage site was not specifically designed to evaluate the natural biodegradation of ethanol, and the monitoring period at both the Sage site and the Florianopolis site is short (14 months and 16 months, respectively). A more refined prediction of the impact of ethanol on BTEX biodegradation requires a better understanding of the fate of ethanol and BTEX following their releases to the subsurface, and a better understanding of the microbial physiology and ecology in the subsurface in the presence of ethanol and BTEX. Thus, the results presented here should be viewed as preliminary, serving as both upper and lower bounds.

Notice

The views expressed in these Proceedings are those of the individual authors and do not necessarily reflect the views and policies of the U.S. Environmental Protection Agency (EPA). Scientists in EPA's Office of Research and Development have prepared the EPA sections, and those sections have been reviewed in accordance with EPA's peer and administrative review policies and approved for presentation and publication.

REFERENCES

Corseuil, H. X, C. S. Hunt, R. C. F. Dos Santos, and P. J. J. Alvarez. 1998. "The Influence of the Gasoline Oxygenate Ethanol on Aerobic and Anaerobic BTX Biodegradation." *Water Research* 32(7): 2065-2072.

Conseuil, H. X., M. Fernandes, M. do Rosario, and P. N. Seabra. 2000. "Results of a Natural Attenuation Field Experiment for an Ethanol-Blended Gasoline Spill."

Proceedings of NGWA/API Conference and Exposition on Petroleum and Organic Chemicals in Ground Water: Prevention, Detection and Remdiation. Anaheim, California, November 15-17, 2000, pp. 24-31.

Hunt, C. S., L. A. Cronkhite, H. X. Corseuil, and P. J. Alvarez. 1997. "Effect of Ethanol on Anaerobic Toluene Degradation in Aquifer Microcosms." In *Division of Environmental Chemistry Preprints of Extended Abstracts.* pp. 424-26. 213th American Chemical Society National Meeting. San Francisco, California: American Chemical Society, April 13-17.

Kennedy, L. G., J. W. Everett, K. J. Ware, R. Parsons and V. Green. 1998. "Iron and Sulfur Mineral Analysis Methods for Natural Attenuation Assessments." *Bioremediation Journal* 2(3&4):259-276.

Ravi, K., J. T. Wilson and L. Dunlap. 2000. "Evaluating Natural Attenuation of MTBE at Multiple UST Sites." *Proceedings of NGWA/API Conference and Exposition on Petroleum and Organic Chemicals in Ground Water: Prevention, Detection and Remdiation.* Anaheim, CA, November 15-17, 2000, pp. 32-49.

Malcom Pirnie, Inc. 1998. *Evaluation of the Fate and Transport of Ethanol in the Environment.* Report prepared for the American Methanol Institute, Washington DC. Malcom Pirnie, Inc., Oakland, CA. 3522-001.

McNab, W., S. E. Heermann and B. Dooher. 1999. "Screening Model Evaluation of the Effects of Ethanol on Benzene Plume Lengths." *Health and Environmental Assessment of the Use of Ethanol as a Fuel Oxygenate.* Chapter 4 of Volume 4. Lawrence Livermore National Laboratory, Livermore, CA. UCRL-AR-135949.

Mravik, S. C., G. W. Sewell and A. L. Wood. 2000. "Co-solvent-based Source Remediation Approach." *Physical and Thermal Technologies: Remediation of Chlorinated and Recalcitrant Compounds.* Battelle Press, Columbus, OH.

Suarez, M. P. and H. S. Rifai. 1999. "Biodegradation Rates for Fuel Hydrocarbons and Chlorinated Solvents in Groundwater." *Bioremediation Journal* 3(4): 337-362.

Suflita, J. and Mormile, M. 1993. "Anaerobic Biodegradation of Known and Potential Gasoline Oxygenates in the Terrestrial Subsurface." *Environmental Science & Technology* 27(5):976-78.

Ulrich, G. 1999. *The Fate and Transport of Ethanol-blended Gasoline in the Environment.* Governor's Ethanol Coalition, Lincoln, Nebraska

Wiedemeier, T. H., H. S. Rifai, C. J. Newell and J. T. Wilson. 1999. *Natural Attenuation of Fuels and Chlorinated Solvents in the Subsurface.* John Wiley & Sons, Inc. New York, NY. pp. 213-216.

IN SITU REMEDIATION OF ETHANOL, AMMONIA, AND PETROLEUM HYDROCARBONS VIA AEROBIC BIODEGRADATION

Mark D. Nelson, P.E., (Delta Environmental Consultants, Inc., St. Paul, MN)
Brad W. Koons, E.I.T.(Leggette, Brashears & Graham, Inc., St. Paul, MN)
Bradley J. Peschong (Leggette, Brashears & Graham, Inc., Sioux Falls, SD)
Carolyn Boben (Williams, Roseville, MN)

Abstract: Soil and ground water at a bulk petroleum storage facility was heavily impacted by a recent release of over 1,000 barrels of ethanol denatured with five percent gasoline. This release occurred in an area previously affected by historical releases of ammonia-based fertilizer and diesel fuel. A variety of microorganisms are capable of metabolizing each of the dissolved components, ethanol, ammonia, and diesel range organics, via multiple enzymatic pathways. However, concentrations of ethanol at this facility, as great as 38,000 milligrams per liter in ground water, have been shown to be inhibitory to soil microorganisms.

Air sparging (AS) was selected as a remedial method to address ethanol impacts to ground water. The intent of the AS system installation was to oxygenate a historically anoxic aquifer in an effort to encourage biological degradation of ethanol. Oxygenation of the subsurface environment stimulated rapid aerobic biodegradation of ethanol. The vigorous bioactivity in the area may have also affected ammonia and petroleum hydrocarbon concentrations in the remedial zone, however, no discernable trends indicating aerobic or anaerobic transformation of these components were found. It is likely that preferential microbial metabolism of ethanol limited electron acceptor availability for nitrification and hydrocarbon uptake.

INTRODUCTION

The study was performed at a petroleum products distribution terminal located in the northern plains. This terminal stores and distributes a variety of petroleum products including gasoline, diesel, and fuel oils. The terminal also stores ethanol denatured with five percent gasoline that is blended with gasoline to produce oxygenate-enhanced fuels.

A denatured ethanol release occurred at the terminal over a 24-hour period in June 2000. It was estimated that 15,000 gallons of ethanol were not recovered by emergency response actions. The lost ethanol formed a pond near the storage tank (Figure 1). Evaporation modeling results indicated that approximately 10,000 gallons of the release were removed through evaporation. It is assumed that the balance of the ethanol infiltrated into the soil.

The area where the release occurred is underlain by a layer of clay from ground surface to between 3 and 4.5 feet below grade surface (bgs). This clay layer was absent at borings RP-4, RP-12, and RP-15, where a medium to fine grain sand was observed from surface to 15 feet bgs. The surficial clay is underlain by fine to very coarse sand, pebbles, and cobbles. The sand layer is underlain by a continuous oxidized till at depths ranging from 15 to 20 feet bgs. This oxidized till is an undulating regional feature observed in previous site assessments at similar depths and is inferred to be a continuous confining layer.

Ground water is encountered at approximately 14 feet bgs. The saturated thickness above the oxidized till is limited to between 18 and 36 inches. Historical petroleum hydrocarbon releases have impacted ground water underlying the terminal. Components of the dissolved plume include benzene, toluene, ethylbenzene, xylenes (BTEX), and diesel range organics. In addition, releases of an ammonia-based fertilizer have created a dissolved ammonia plume that is commingled with the hydrocarbon plume. These dissolved hydrocarbons and ammonia have exerted a substantial oxygen demand. Site data indicate that the ground water is generally anoxic with observed elevated methane concentrations.

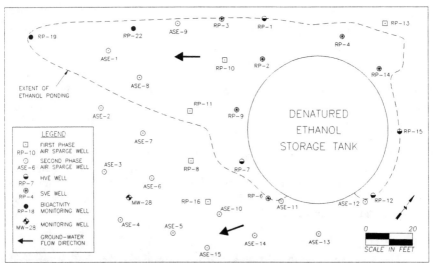

FIGURE 1. Site map detailing the area where the ethanol pond formed, wells used for site remediation and observation, and the direction of ground-water flow.

SITE INVESTIGATION

Twenty-two soil borings (RP-1 through 22) were installed in the release area to delineate the extent of ethanol impacts to the soil and ground water (Figure 1). The borings were advanced to 25 feet below grade. Several of the borings were completed as monitoring wells with ten-foot screened intervals extending from the base of the oxidized clay layer to between four and five feet bgs. Soil samples recovered from auger spoils and/or from split spoon samples were field screened with a photoionization detector. Additional samples were collected for laboratory analysis of ethanol.

Field screening results from most borings exhibited ethanol impacts in the upper five feet of soil. RP-12 and RP-15 were exceptions; these borings appeared to be impacted from the surface to the water table at approximately 14 ft bgs. Near the water table elevated concentrations of organic compounds were measured in each of the borings, however, odors associated with these samples indicated that the measured concentrations were likely diesel fuels.

Laboratory analysis of the soil samples confirmed the results of the field screening analysis. Surface soils at RP-6, 7, 9 through 13, 15, 19 and 22 were significantly impacted by ethanol with concentrations ranging from 2,016 to 47,405 milligrams per kilogram. Ethanol was detected at the ground-water table interface in RP-15, but in no other boring. Analysis of ground water collected three days after the second release indicated that ethanol was present in MW-28 at a concentration of 25 milligrams per liter (mg/L) and no ethanol was present in samples collected from RP-10 or RP-15. Because ethanol was not detected at MW-28 in two subsequent sampling events and the ethanol pond limits did not extend into the vicinity of MW-28, it is believed that this initial ethanol detection was an analytical error.

REMEDIAL RESPONSE

Multiple remedial systems were installed at the site in response to the ethanol release. These systems were installed in two phases. The first phase was completed within two weeks of the release and consisted of a soil-vapor extraction (SVE) system, a high-vacuum extraction (HVE) system, and an air-sparging (AS) system. The SVE and HVE systems were operated for a short duration prior to the onset of cold weather. The AS system utilized a 55 standard cubic feet per minute (scfm) blower operating at a pressure of three pounds per square inch (psi) and injected atmospheric air into five two-inch diameter wells (Figure 1).

Increased ethanol concentrations in the wells in, and immediately down gradient of the release area prompted a more aggressive remedial strategy that would not be impeded by cold weather. Additional air sparging was selected as the most efficient remedial response. The second AS system installed within the release area utilized a 110-scfm blower operating at 14 psi and injected atmospheric air into 15 one-inch diameter wells. All AS well screens were one foot in length and were located approximately three feet below the ground-water table. The first and second systems began operation in late June and early October 2000, respectively.

A sampling program was initiated to monitor the effect of the remediation efforts on ground-water quality. Several wells were monitored periodically for ethanol, dissolved oxygen (DO), temperature, dissolved ammonia, nitrate, nitrite, BTEX, gasoline range organics, and diesel range organics.

RESULTS

A decrease in ethanol concentrations was observed in all wells where ground-water samples were collected (Figure 2). Ethanol concentrations in MW-28 and RP-11 rapidly decreased from 1,689 and 2,410 mg/L, respectively, to below the laboratory method detection limit of 5 mg/L in a four to six week period. Slight increases in ethanol concentrations were observed over a two-month period in RP-6 and RP-22 from 34 to 38 mg/L and 87 to 132 mg/L, respectively. These marginal increases were followed six weeks later by concentrations that were below detection limits for ethanol. The most marked decrease in ethanol concentration was observed at RP-2. Between late August 2000, and early January 2001, the concentration of ethanol in RP-2 declined from 38,076 to 13.3 mg/L, a three orders of magnitude decrease.

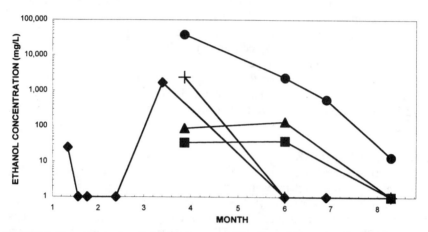

FIGURE 2. Logarithmic plot of ethanol concentrations in wells MW-28, RP-2, RP-6, RP-11, and RP-22. A reduction in ethanol concentration was observed in all wells. Symbols are as follows: ◆ – MW-28; ● - RP-2; ■ - RP-6; + - RP-11; ▲ – RP-22.

Dissolved oxygen data acquired in early October 2000, indicated that ground water in the release area was anoxic. This result was expected as earlier assessments reported anoxic conditions in the ground water across the site. Dissolved oxygen in well RP-11 increased from less than 1 to 4.7 mg/L during an early November sampling event, and increased again to 10.5 mg/L in early December. Because RP-11 is used as an AS injection point, an increase in DO was anticipated, however, an increase was not detected until four months after the AS system began operating. No DO data was collected from other AS points, but it is likely that similar trends occurred. An increase in DO was also observed in MW-28 during a mid February 2001, sampling event. The DO concentration in this well increased from a previous average over three sampling events of 0.4 mg/L to 1.9 mg/L. No significant change in DO concentrations was noted in any of the other monitored wells.

Elevated ground-water temperatures were observed in wells RP-2, RP-19, and RP-22 (Figure 3). Between October and December 2000, ground-water temperatures in these wells increased at a similar rate. Following the December 2000 sampling event, ground-water temperatures at RP-2 and RP-19 decreased, conversely, ground-water temperature at RP-22 continued to increase. Ground-water temperatures at wells MW-28, RP-12, and RP-15 decreased throughout the same period.

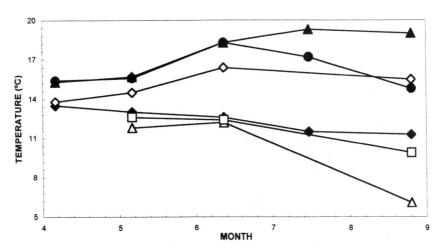

FIGURE 3. Elevated ground-water temperatures were noted in wells RP-2, RP-19 and RP-22, all of which were located within the ethanol pond area. Symbols are as follows: ♦ – MW-28; ● - RP-2; △ – RP-12; □ - RP-15; ◇ - RP-19; ▲ – RP-22.

Concentrations of petroleum hydrocarbons in monitored wells varied with no discernable trend. No correlation between ethanol concentrations and petroleum hydrocarbons was observed. Similarly, no correlation between dissolved forms of nitrogen in wells MW-28, RP-2, RP-6, and RP-11 could be distinguished. Ammonia concentrations in MW-28 and RP-6 decreased over the sampling period and increased in RP-2 and RP-11. Concentrations of nitrate increased in wells MW-28 and RP-11, and no clear trend was observed in RP-2 or RP-6. Concentrations of nitrite between 2.8 and 3.2 mg/L were observed in ground water taken from RP-6, but were not observed in samples collected from other wells.

DISCUSSION

Little literature is available concerning the transport of bulk quantities of ethanol through the subsurface. The data presented here was not collected with the intent of providing a rigorous evaluation of ethanol transport, however, an observation of interest can be made. The initial subsurface investigation found no ethanol in ground-water samples and limited ethanol impacts to soils below five feet bgs, yet August sampling results demonstrated significant ethanol concentrations in multiple wells. A transport mechanism for ethanol moving through the vadose zone cannot be verified due to the lack of available data, however, given the speed and extent of the impacts to the ground water, it appears that transport was likely through secondary porosity features of the surficial clay layer. It is likely that rainwater accumulating in the same area as the ethanol pond flushed large amounts of ethanol into the ground water.

Concentrations of ethanol in the aquifer rapidly decreased over the period of the study. It is likely that this decrease was due to subsurface bioactivity, which resulted in increased ground-water temperatures. Background groundwater temperatures, up gradient of the spill area, averaged approximately 11 degrees centigrade (°C). The average temperature at RP-22, on the down gradient edge of the release, was 17.5 °C over a similar time period. This significant temperature increase through the release area, coupled with the decreasing ethanol concentration trend, indicates that biodegradation had occurred.

Microbial degradation of ethanol was expected; studies have reported that ethanol is readily consumed by a variety of microorganisms (Alvarez, 1999; Ulrich, 1999). High concentrations of ethanol, as in the case of a denatured ethanol release, are a biodegradation concern as ethanol is an effective disinfectant. Concentrations in excess of 40,000 mg/L in water can be toxic to microorganisms and inhibition of microbial activity has been shown to occur at concentrations over 10,000 mg/L (Hunt, 1997). Examination of the ethanol concentration and correlating ground-water temperature data for RP-2 (Figure 4) indicates that such an inhibition occurred. A rapid initial decrease in ethanol concentration from 38,076 to 2,330 mg/L occurred between late August and early November 2000, and was likely the result of abiotic transport mechanisms. Following this concentration decrease, the rate of ground-water temperature ascension increased. This rate increase indicates that biological activity was more vigorous at the lower ethanol concentration.

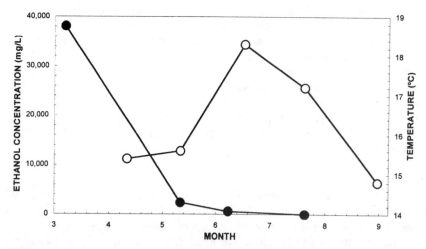

FIGURE 4. Ethanol concentration and ground-water temperature at RP-2. A reversal of RP-2 ground-water temperature increases corresponded with reduced ethanol concentrations at that location. Symbols are as follows: ● - ethanol concentration in ground water recovered from RP-2; ○ - ground-water temperature at RP-2.

No clear trends were identified for oxidation of ammonia or degradation of petroleum hydrocarbons. Nitrification of ammonia was expected in the remedial zone. It was also anticipated that the biologically produced nitrate would be utilized as an electron acceptor in down gradient anaerobic degradation of ethanol or petroleum hydrocarbons. Because no significant decreases in petroleum hydrocarbon or ammonia concentrations were observed, it is hypothesized that all introduced oxygen was coupled to microbial uptake of ethanol. It is hoped that future monitoring data will demonstrate a shift from ethanol degradation to nitrification and petroleum hydrocarbon uptake as ethanol concentrations in the aquifer approach zero.

CONCLUSIONS

Evaluation of the data generated from this site allow the following generalized conclusions:

- Transport of ethanol through the vadose zone and into ground water can be rapid. It is important that the environmental professional understand that there is little or no retardation and, as a result, swift movement of ethanol plumes can be expected.

- There was no indication of near term co-solvent related increases in dissolved hydrocarbon concentration within the existing plume. Long term monitoring of the site may provide additional information regarding co-solvent effects.

- Biodegradation of ethanol can be a successful remedial approach. High concentrations of ethanol may be inhibitory, however, application of air sparging or similar oxygen introduction technologies can be successful in treating impacts resulting from ethanol releases.

- Ethanol is preferentially degraded in aerobic environments when commingled with hydrocarbon plumes. It is likely that dissolved oxygen will quickly be consumed in ethanol-impacted ground water.

REFERENCES

Alvarez, P. J. J., and C. S. Hunt. 1999. "The Effect of Ethanol on BTEX Biodegradation and Natural Attenuation." In *Health and Environmental Assessment of the Use of Ethanol as a Fuel Oxygenate: Report to the California Environmental Council in Response to Executive Order D-5-99.* Volume 4, chapter 3, pp. 3-1 – 3-37. Lawrence Livermore National Laboratory Publication UCRL-AR-135949, Livermore, CA. .

Hunt, C. S. and P. J. J. Alvarez. 1997. "Effect of Ethanol on Aerobic BTX Degradation." *In Situ and On-Site Bioremediation: Papers from the Fourth International In Situ and On-Site Bioremediation Symposium.* Volume 1, pp. 49-54. Battelle Press, Columbus, OH.

Ulrich, G. 1999. *The Fate and Transport of Ethanol-Blended Gasoline in the Environment: A Literature Review and Transport Modeling.* Governor's Ethanol Coalition, Lincoln, NE.

METHANOL BEHAVIOR IN THE SUBSURFACE AT A COASTAL PLAIN RELEASE SITE

H. James Reisinger (Integrated Science & Technology, Inc., Marietta, Georgia)
Julie B. Raming, (Georgia-Pacific Corporation, Atlanta, Georgia)
Adam J. Hayes (Triple Point Engineers, Inc., Roswell, Georgia)

ABSTRACT: In 1991, neat methanol was discovered in the subsurface at a chemical manufacturing facility in coastal South Carolina. Site characterization showed that methanol from a leaking pipe had infiltrated through the 15-ft thick silty fine-grained sand vadose zone and, being slightly less dense than water, accumulated on the ground-water surface as a short-lived separate phase. With the passage of time, a dissolved-phase plume developed through advective dispersive transport. Following detailed feasibility testing, it was determined that monitored natural attenuation (MNA) was best suited to address the dissolved-phase methanol plume. Over an extended monitoring period maximum plume methanol concentrations decreased from about 900,000 to about 340,000 mg/L and the plume area decreased 45 percent. Monitoring results indicated that the aquifer was anaerobic and nearly devoid of nitrate and sulfate. Deeply reducing conditions and the accumulation of up to 85 percent methane in the vadose zone atmosphere demonstrated that the methanol was being biodegraded via methanogenesis. A soil-vapor extraction (SVE) system was developed to remove the methane and to attempt to enhance the biodegradation rate. Data generated through pilot testing conservatively showed that the estimated conservative methane production rate was 3 percent per day, which can be directly related to the methanogenic biodegradation rate.

INTRODUCTION

In 1991, neat methanol was discovered erupting through asphalt paving at a chemical manufacturing facility in coastal South Carolina. The source of the methanol was piping that conveyed it from an aboveground storage tank system to the main manufacturing area. The methanol piping was replaced with an aboveground conveyance system when the leak was discovered. At the time of discovery, the release was reported to the responsible regulatory agency and assessment activities were initiated to ascertain the extent and distribution of methanol in the subsurface.

The site is set in the South Carolina Coastal Plain at an elevation of about 23 m NVGD and has relatively low relief. The soil underlying the site is primarily sand and silt with some clay, and it exists as an upward fining sequence. Ground water occurs under water table conditions at a depth of 4.6 to 6.1 m below ground surface. Ground-water flow is predominately to the south-southeast at a gradient of 0.01 m/m and the average hydraulic conductivity is 9.42 m/day, which is consistent with the soil character.

Site characterization was accomplished using a combination of conventional and rapid assessment techniques. The resultant data showed minimal soil impact. Results of ground-water sampling and analysis showed that a plume of separate phase and dissolved-phase methanol existed beneath the site. The plume extended from the release site to an area just upgradient of a ditch that runs along the eastern property boundary (Figure 1). The dissolved-phase concentration in the immediate vicinity of the release site as well as gauging observations demonstrated the presence of a separate-phase plume. The analytical data also showed that the concentrations decreased both longitudinally (in the direction of ground-water flow) and transversely (in the direction perpendicular to ground-water flow). In addition, the data indicated that the dissolved-phase plume was about 7.6-m thick. The estimated methanol volume was in the range of 1,893 to 3,785 m^3. However, this estimate must be viewed with caution as it is based on numerous assumptions and interpolation of data among the monitor well network that exists on the site.

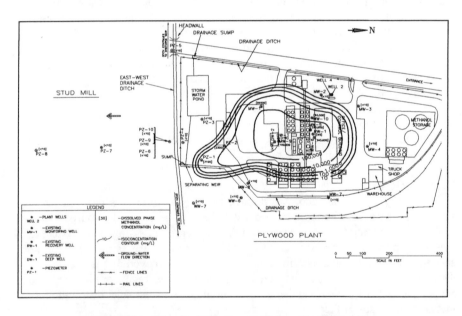

FIGURE 1. Site and Methanol Plume Configuration

After site characterization was completed, a monitoring program was implemented, and efforts were initiated to identify the most prudent approach to site remediation, which was required by the responsible regulatory agency. The Agency, apparently required remediation as a result of the large volume released and the fact that methanol is a Class B poison. Remediation options considered included ground-water extraction and treatment with an aboveground biological reactor, ground-water extraction and treatment via aeration and biodegradation in an existing treatment system, and MNA. During the feasibility evaluation (including pumping tests and pilot studies), ground-water quality monitoring continued. Results of feasibility testing and evaluation showed that all of the options considered were feasible and could be applied at the site. However, monitoring data showed that dissolved-phase methanol plume migration had stabilized and that the mass was decreasing. It was therefore determined that the most appropriate approach to site remediation was MNA.

With the passage of time, the estimated methanol mass decreased about 45 percent and the dissolved-phase plume area decreased concomitantly. However, given the sizable initial mass, a protracted time would be required to achieve site closure. Therefore, efforts were initiated to develop an approach that more quickly reduced the methanol mass in the source. Among the options considered were ground-water extraction and treatment, and air sparging. After due consideration, air sparging to stimulate *in situ* methanol biodegradation was selected. A synoptic ground-water sampling and analysis event was completed to prepare for pilot testing. Higher-than-normal pressure was observed in several of the monitoring wells in the highest concentration aspects of the dissolved-methanol plume (as high as 34 kPa). It was assumed that the pressure resulted from biogenic gas production and samples were collected and analyzed to confirm this and to determine the gas composition. Analysis results showed that the headspace vapor was about 65 percent methane, which suggested methanogenic methanol degradation. This was consistent with the dissolved-oxygen and secondary electron acceptor distribution as well as the oxidation-reduction potential.

Discovery of vadose zone methane concentrations that exceeded the upper explosive limit and oxygen concentrations less than that needed to support combustion or explosion prompted a decision to forgo oxygen enrichment through air injection. A vadose zone vapor survey was conducted to determine the methane distribution; the results showed that the highest methane concentrations (about 85 percent) corresponded to those areas where the dissolved-phase methanol concentrations were highest. Analysis of the collective database suggested that a combination of methanogenic MNA and SVE to reduce the methane concentration to safe levels would be the optimal remediation approach for the site. A pilot test was conducted to develop the data needed to design a system.

METHANOL CHARACTER, TRANSPORT, AND FATE

Methanol is a clear colorless liquid with a molecular weight of 32.04 grams/mole and a boiling point of 65°C. Fate and transport of dissolved-phase methanol are primarily functions of its specific physicochemical characteristics. Although methanol has a vapor pressure of 92-mm Hg at 20°C, it is also miscible in water. Therefore, volatilization of dissolved methanol will be negligible, as its Henry's Law constant of 0.00695 atm-L/mol suggests. Another common fate mechanism for dissolved-phase compounds is partitioning into organic matter in the aquifer sediments. However, methanol has a log octanol/water partitioning coefficient (K_{ow}) of -0.74 ± 0.1. This indicates that methanol tends to remain in the dissolved phase. These factors as well as its relatively small molecular size indicated that methanol should travel at approximately the same rate as groundwater flow.

Based on the relative insignificance of physicochemical attenuation processes and the lack of dissolved-phase plume growth over several years, intrinsic biological degradation was suspected as the primary mechanism controlling the plume. Dissolved-phase methanol may be degraded aerobically or anaerobically by indigenous heterotrophic bacteria. These microorganisms are capable of totally mineralizing methanol *in situ*. The end products of aerobic degradation are carbon dioxide and water, and the end products of anaerobic degradation typically include carbon dioxide and water, as well as reduced species of the inorganic terminal electron acceptor. Methane is the end product of methanogenic methanol degradation. The following equations depict half redox reactions and complete methanol mineralization reactions for the most common aerobic and anaerobic pathways.

Half Redox Reactions

OXYGEN (O_2) (1)
$$O_2 + 4 H^+ + 4 e^- \rightarrow 2 H_2O$$

NITRATE (NO_3^-) (2)
$$NO_3^- + 6 H^+ + 5 e^- \rightarrow \tfrac{1}{2} N_2 + 3 H_2O$$

SULFATE (SO_4^{-2}) (3)
$$SO_4^{-2} + 9 H^+ + 8 e^- \rightarrow HS^- + 4 H_2O$$

IRON III ($Fe(OH)_3$) (4)
$$Fe(OH)_3 + 3 H^+ + e^- \rightarrow Fe^{2+} + 3 H_2O$$

Methanol Oxidation Reactions

OXYGEN (O_2) (5)
$$CH_3OH + \tfrac{3}{2} O_2 \rightarrow CO_2 + 2 H_2O$$

NITRATE (NO_3^-) (6)
$$CH_3OH + \tfrac{6}{5} NO_3^- + \tfrac{6}{5} H^+ \rightarrow CO_2 + \tfrac{3}{5} N_2 + \tfrac{13}{5} H_2O$$

SULFATE (SO_4^{-2}) (7)
$$CH_3OH + \tfrac{3}{4} SO_4^{2-} + \tfrac{3}{4} H^+ \rightarrow \tfrac{3}{4} HS^- + CO_2 + 2 H_2O$$

IRON III ($Fe(OH)_3$) (8)
$$CH_3OH + 6 Fe(OH)_{3(s)} + 12 H^+ \rightarrow 6 Fe^{2+} + CO_2 + 17 H_2O$$

Fermentative/Methanogenic Degradation (9)

$$CH_3OH + H_2O \rightarrow CO_2 + 6H^+$$

$$3CH_3OH + 6H^+ \rightarrow 3CH_4 + 3H_2O$$

$$4CH_3OH \rightarrow 3CH_4 + CO_2 + 2H_2O$$

This expression from White (1986) is a simplification of the process and does not consider methanotrophic processes that serve as sinks for produced methane.

When comparing the aerobic and anaerobic biodegradation half-lives of methanol to those of readily degraded petroleum constituents such as benzene and toluene, methanol proves more easily biologically degraded than these hydrocarbons. The aerobic and anaerobic aqueous biodegradation half-lives for methanol are 7 and 5 days, respectively. The corresponding aerobic and anaerobic half-lives for benzene are 16 days and up to 24 months. Likewise, aerobic and anaerobic aqueous biodegradation half-lives for toluene are 22 days and 30 weeks, respectively (Howard et al. 1991). This comparison shows methanol is highly biodegradable and would be expected to be short lived in the environment.

However, under some circumstances, such as those at the South Carolina site, methanol can be more persistent. Methanol, like most alcohols, can be antiseptic at high concentrations. That is, it can be bactericidal or bacteriostatic. Malcolm Pirnie (1999) reported that methanol was toxic to microorganisms in the subsurface at concentrations greater than 100,000 mg/L. It is likely that at least a portion of the toxicity is a function of methanol's high aqueous solubility and resultant high bioavailability. The highest concentration region of the South Carolina plume is thus expected to have greatly reduced or completely inhibited biological activity.

The physicochemical characteristics and fate processes outlined in this section dictate methanol's behavior in the subsurface. The miscibility and nearly non-existent sorbtion potential render methanol and other alcohols extremely mobile in the subsurface. When released to the environment, neat methanol mixes completely in the water. When part of a mixture such as gasoline, methanol and other alcohols partition into water proportional to their mole fraction in the mixture thereby reducing the mass transfer relative to neat methanol. Once dissolved in ground water, methanol and other alcohols are transported with the ground water at essentially the same rate as the ground water. Furthermore, the size and polar character of the methanol molecule affect the degree to which diffusion and dispersion affect methanol transport.

In general, biodegradation plays a major role in the ultimate fate of xenobiotics in the environment. This is particularly true of methanol and other alcohols. As suggested in this section, methanol is highly biodegradable, being mineralized through aerobic or anaerobic pathways. The product of the methanol and other alcohol fate and transport processes would generally be expected to be a plume that stabilizes relatively rapidly and is ultimately eliminated relatively rapidly.

SOUTH CAROLINA METHANOL PLUME BEHAVIOR

The South Carolina methanol plume appears to be somewhat unique in light of the theoretically expected behavior and fate of an alcohol plume. Rather than develop, quickly stabilize, decay, and ultimately extinguish, the South Carolina plume developed, migrated about 305 m downgradient, and grew to occupy an area of about 18,600 m^2. It is difficult to estimate the time to stabilization because the release date is unknown. However, at the time of site characterization, the plume had apparently achieved stability. Figure 2 shows the evolution of the plume over the past 9 years; Figure 3 shows the dissolved- phase methanol temporal trends in two of the most impacted monitor wells (MW-6 and MW-9) and two of the monitor wells in the more downgradient aspect of the plume (PZ-1 and PZ-2).

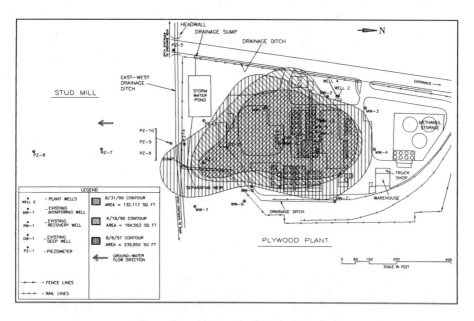

FIGURE 2. Methanol Plume Evolution

These figures show that the source area concentrations were highly variable with a distinct downward trend over the monitoring period. However, in addition to methanogenic biodegradation, a sizable portion of the decrease can be attributed to downgradient advective transport as well as diffusive and dispersive transport. Figure 2 shows that the South Carolina plume is quite extensive laterally, a function of the relatively flat gradient and the more dominant effect of diffusion and dispersion as fate mechanisms. In contrast, the downgradient concentrations not only decreased, but the leading edge of the plume, and therefore the plume area, decreased. It appears that because the leading edge of the plume is a transition zone between oxidizing and reducing conditions methanol is being consumed at least at the rate at which it is being transported in the downgradient direction, thus, achieving plume stability.

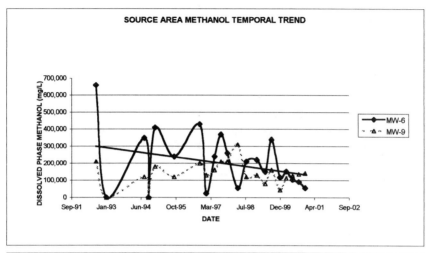

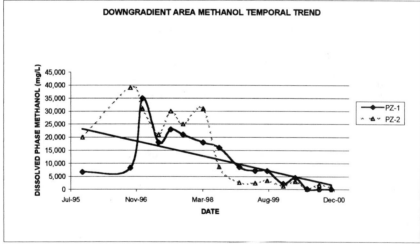

FIGURE 3. Methanol Temporal Trends

Evidence that the South Carolina methanol plume is stable or decreasing is apparent in the concentrations and distributions of parameters indicative of natural attenuation control. This site more or less fits a natural attenuation model. Its plume morphology clearly shows the effects of both transverse and longitudinal dispersion because it is generally elongate with a strong lateral component. Electron donor concentrations and mass are relatively high and decline in the downgradient direction. Electron acceptor distribution follows the classical pattern.

The highest methanol concentration aspect of the plume is essentially devoid of oxygen, nitrate-nitrogen, and nitrite-nitrogen, but this aspect exhibits substantially higher ammonia-nitrogen concentrations. Sulfate concentrations are lower within the high methanol concentration aspect of the plume but are still detectable. Substantial iron concentrations exist in the methanol plume. These are dominated by the ferrous species, a reflection of ferric iron reduction and strongly reducing conditions. Alkalinity is relatively high throughout the plume. A large portion of the plume also exists under reducing conditions.

As described previously, the vadose zone above the plume contains very high methane concentrations, which, along with the strongly reducing conditions, is evidence of substantial methanogenic methanol biodegradation. Obviously, the substantial methane accumulation observed at this site is somewhat different than typically observed at sites impacted with petroleum hydrocarbons. This is a function of the high oxygen demand exerted by the highly biodegradable methanol and the methane that is a primary product of anaerobic methanol biodegradation.

METHANE PRODUCTION KINETICS AT THE SOUTH CAROLINA SITE

Unlike many other sites (i.e., hydrocarbon release sites) at which methanogenic biodegradation is occurring, the South Carolina site has a substantial volume of methane that has accumulated in the vadose zone atmosphere. Observed methane concentrations exceed the upper explosive limit over a large aspect of the dissolved-phase methanol plume. It appears that the subsurface structure, the partially paved surface character, and the large methanol source in the subsurface contribute to the methane accumulation. At more typical sites, the methanol or other alcohol source is substantially smaller and the volume or mass of methane produced is proportionately smaller. In fact, it is likely that the smaller methane mass diffuses to the atmosphere or is consumed by methanotrophic microorganisms at a rate that precludes accumulation.

At the South Carolina site, the methane concentrations created a potential safety hazard. Therefore, site remediation addressed accumulated vadose zone methane as well as dissolved-phase methanol. Because the dissolved-phase plume was stable or decaying and because there were no potential off-site receptors, MNA was selected to address the dissolved phase. SVE was selected to address the vadose zone methane. It has also been hypothesized, based on thermodynamics analysis, that removal of the methane may increase the rate of methanogenic methanol biodegradation. Calculation of Gibbs free energy for the reaction suggests that at very high methane concentrations, methanogenesis may become thermodynamically unfavorable. While this would, under most circumstances, seem to have a low probability as a result of methane's low solubility in water, it appears that at the South Carolina site, methane accumulation in the vadose zone to very high levels (i.e. up to 85%) has indeed resulted in aqueous supersaturation as evidenced by monitor well pressurization. This strongly suggests that methane could be accumulating to potentially inhibitory levels in the saturated zone and that methane removal could render the environment more thermodynamically favorable for methanogenesis.

A pilot test was conducted to generate the data needed to design a SVE system. The pilot test was conducted by applying a vacuum to extraction wells. An electric regenerative blower supplied vacuum. During the test, vadose zone atmospheric flow, temperature, pressure, and methane were monitored in the system off-gas. Vacuum and air flow were measured in monitoring points surrounding the extraction points. The test was run for 34 days; 17 days of extraction and 17 days of post-extraction monitoring. During extraction, the methane concentrations were reduced from about 50 percent to non-detectable levels, thereby, demonstrating the efficacy of SVE as a site remediation tool. The post-extraction monitoring was conducted to assist in estimating the rate at which methane accumulated in the subsurface. It was assumed that methane accumulation in the vadose zone was a representation of the methane production rate. The following figure (Figure 4) shows plots of methane rebound in three of the points monitored post injection.

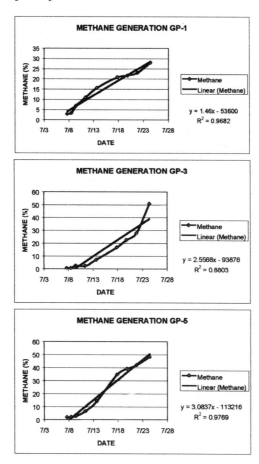

FIGURE 4. Methane Production Rate Analysis

The relationships show a nearly linear relationship between methane concentration (percent) and time. Regression analysis and conversion of the rates to mass removed per unit time shows an arithmetic mean methane generation rate of 0.0124 moles per day with a range of 0.003 to 0.031 moles per day (n=9), which can reasonably be assumed to represent a conservatively estimated arithmetic mean methanol biodegradation rate of 0.0165 moles per day with a range of 0.004 to 0.0414 moles per day (n=9). The estimated observed rate is conservative as it does not consider methanotrophic processes through which produced methane could be lost. However, given that extraction was not performed throughout the site, it is possible that a portion of the accumulated methane could also be attributed to diffusion from the higher concentration areas that surround the area where extraction was conducted.

SUMMARY AND CONCLUSIONS

The fact that methanol and other alcohols are readily biodegradable is well established. Alcohols are often used as a carbon source in wastewater treatment. While alcohols provide carbon for cell growth and energy, they also exert a significant oxygen demand. Therefore, it follows that alcohols released into the subsurface should be readily biodegradable and be relatively short lived as suggested by Malcolm Pirnie (1999). However, biodegradability does not necessarily translate to a total lack of plume persistence. In fact, at the South Carolina site, the plume has persisted for 9 years. It is likely that the plume has endured so long due to its large initial mass. The mass may have taxed the system by depleting electron acceptors and inhibiting or killing the indigenous microbial consortium. In concert, these have allowed the plume to persist.

It is notable that a substantial portion of the initial methanol mass introduced to the environment at the South Carolina site has been biodegraded, most likely through methanogenic and oxidative reactions. Results of methane production rate estimates, which are assumed to be directly related to methanol biodegradation rates, are relatively high despite or as a result of the persistence of a large residual methanol mass. Thus, it is important to evaluate each alcohol release site as a unique entity and to resist the temptation to assume that because alcohols are traditionally short lived in the environment, they will be short lived at each site.

REFERENCES

Howard, Philip H, Robert S. Boethling, William F. Jarvis, William M. Meylan, and Edward M. Michalenko. 1991. *Handbook of Environmental Degradation Rates*. Lewis Publishers, Chelsea, MI.

Malcolm Pirnie, Inc. 1999. *Evaluation of the Fate and Transport of Methanol into the Environment*. Prepared for American Methanol Institute by Malcolm Pirnie, Inc. American Methanol Institute, Washington, DC.

White, Kevin Duane. 1986. "A Comparison of Subsurface Biodegradation Rates of Methanol and Tertiary Butanol in Contaminated and Uncontaminated Sites."

Ph.D. Dissertation, Virginia Polytechnic Institute and State University, Blacksburg, VA.

EFFECTS OF ETHANOL VERSUS MTBE ON BTEX NATURAL ATTENUATION

Marcio L.B. da Silva[1], Graciela M. Ruiz[1], Jose M. Fernandez[1], Harry R. Beller[2], and Pedro J.J. Alvarez[1]
[1]University of Iowa, Civil and Environmental Engineering, Iowa City, IA, and
[2]Lawrence Livermore National Laboratory, Livermore, CA

ABSTRACT: The increased use of ethanol (EtOH) as replacement for the gasoline oxygenate, methyl *tert*-butyl ether (MtBE), may lead to indirect impacts related to natural attenuation of BTEX compounds. In theory, both MtBE and EtOH could enhance dissolved BTEX mobility by exerting a cosolvent effect that decreases sorption-related retardation. Such effects, however, are concentration-dependent and were not observed in this work when MtBE or EtOH (at 1% v/v) was added continuously with BTEX to sterile aquifer columns. Nevertheless, a significant co-solvent effect was observed with 50% EtOH, suggesting that neat EtOH spills in bulk terminals could facilitate the migration of pre-existing contamination. MtBE was not degraded or retarded in this work, and it did not affect BTEX degradation. EtOH, on the other hand, was degraded rapidly and exerted a high demand for nutrients and electron acceptors that could otherwise have been used for BTEX degradation. Microbial depletion of oxygen in EtOH-fed columns caused anaerobic conditions that hindered benzene degradation. EtOH was also the first compound to degrade in aquifer microcosms under different electron acceptor conditions, and its presence had a variable effect on toluene degradation as a function of electron-accepting conditions and bacterial source. In some cases, EtOH retarded toluene degradation, but it occasionally enhanced toluene degradation, presumably due to a low initial concentration of BTEX degraders and their incidental growth during EtOH degradation. The preferential degradation of EtOH by indigenous microorganisms and the accompanying depletion of oxygen and other electron acceptors suggest that EtOH could hinder BTEX natural attenuation and increase plume length.

INTRODUCTION

Monoaromatic hydrocarbons such as benzene, toluene, ethylbenzene, and the three isomers of xylene (BTEX) are ubiquitous groundwater pollutants commonly associated with petroleum product releases. Understanding the factors that affect the fate and transport of BTEX compounds in aquifers is of paramount importance for risk assessment and corrective action purposes. Little attention has been given to how differences in gasoline formulation affect natural attenuation. In this regard, there is a recent initiative being considered to phase out methyl *tert*-butyl ether (MtBE) as a gasoline oxygenate (Federal Register, 2000), due to its recalcitrance, ability to rapidly impact drinking water sources, and low taste and odor thresholds. The most likely candidate to replace MtBE (which accounts for 80% of current oxygenate use) is ethanol (EtOH) (currently accounting for

15% of oxygenate use). EtOH is also a renewable resource that can serve as a substitute fuel for imported oil. Therefore, an increase in the use of EtOH as a gasoline additive seems imminent, and a better understanding of its effects on BTEX migration and natural attenuation is warranted.

In this work, aquifer columns were used to evaluate the potential for EtOH to enhance BTEX migration by decreasing (sorption-related) retardation, and to study biodegradation and geochemical transitions in a flow-through system simulating natural attenuation. Aquifer microcosm experiments were also conducted with different aquifer materials under different electron acceptor conditions to investigate response variability as a function of site specificity.

MATERIALS AND METHODS

Retardation Factor Experiments. Small aquifer columns (Kontes, 15-cm long, 1-cm I.D.) were used to compare the effects of EtOH and MtBE on BTEX retardation factors. Three glass columns were packed with uncontaminated soil (f_{oc} = 0.024) and fed continuously with benzene (5 mg/L), toluene (5 mg/L), and o-xylene (3 mg/L) at a constant flow rate of 1 mL/hr using a Harvard syringe pump. Two columns were also fed with EtOH or MtBE at 10,000 mg/L each, which is a concentration unlikely to be exceeded at sites contaminated with reformulated gasoline. EtOH at 50% (v/v) was also tested to investigate the effect of a neat EtOH spill at a bulk terminal. All columns were poisoned with biocide Kathon CG/IPC (diluted 1:100). Samples were taken every hour and analyzed for BTX, MtBE and EtOH concentrations using a HP 5890 gas chromatograph, as described elsewhere (Alvarez et al., 1998). A carbonate solution (1 g/L as $CaCO_3$) spiked with 1.5 g/L of potassium bromide was used as a tracer to determine the hydraulic characteristics of the columns (Table 1).

TABLE 1: Hydraulic characteristics of small aquifer columns.

Parameter	Value
Porosity (fraction of total volume)	0.45
Darcy velocity, q (cm/hr)	2.8
Dispersion coefficient, D (cm^2/hr)	0.7
Flow rate, Q (mL/hr)	1.0
Soil Organic Fraction (f_{oc})	0.024
Peclet Number (vL/D)	63

Natural Attenuation Experiments. Large aquifer columns (120-cm long, 5-cm I.D., equipped with 8 sampling ports) were used to investigate natural attenuation of benzene, toluene, EtOH and MtBE, and their potential interactive effects. The columns were packed with non-contaminated aquifer material (f_{oc} = 0.01) from a Northwest Terminal, and were kept in the dark at 22°C. Five columns were operated. One column was fed benzene and toluene (BT) at 1 mg/L total. This column served as a baseline for the effect of MtBE or EtOH on BT attenuation. Three other columns were fed BT plus either EtOH (~2 g/L), MtBE (25 mg/L), or both. A sterile control column was also run to discriminate biodegradation from potential abiotic losses. This column was poisoned with Kathon biocide and fed

BT plus EtOH and MtBE. Each column was fed continuously with bicarbonate-buffered synthetic groundwater (van Gunten and Zobrist, 1993) (Table 2) purged with 5% CO_2. Columns were fed continuously in an upflow mode at ~7 mL/h using both a peristaltic pump (Masterflex Mod. 7519-15) to feed the medium and a syringe pump (Harvard Apparatus Mod. 22) to feed BT, EtOH, and MtBE.

TABLE 2: Amendments to the influent of large columns.

Compound	Influent Concentration (mg/L)	
	BT alone	BT plus EtOH or MtBE
Electron Acceptors and Nutrients		
Oxygen (O_2)	8	8
Nitrate (NO_3^-)	30	30
Sulfate (SO_4^{-2})	150	150
Carbonate (as $CaCO_3$)	1,000	1,000
Mg^{2+}	1.5	1.5
PO_4^{-3}	0.06	0.06
Ni(II),Cu(II), Zn(II),Co(II), Mo(IV)	0.002 each	0.002 each
Electron Donors		
Total BT	~1	
EtOH	~2000 or none	
MtBE	~ 25 or none	

The ratio of the peristaltic to syringe pump rates was set at 1:20. Bromide tracer results indicated an effective porosity of about 0.3. The pore velocity in the packed columns was 2 cm/h. The dispersion coefficient was estimated to be 5 cm^2/h. Approximately 6 days were required to displace one pore volume.

1-mL samples from the influent and side ports were collected with gas-tight syringes (Hamilton Co., Reno, Nev.). BT, EtOH, MtBE and methane were analyzed using an HP 5890 gas chromatograph, as described elsewhere (Alvarez *et al.*, 1998). Bromide, acetate, nitrate and sulfate were analyzed in an Alcott 728 autosampling ion chromatograph interfaced to a conductivity detector (Dionex Inc., Sunnyvale, CA). The pH of the samples was measured using a Fisher Scientific AB15 pH meter. Oxidation-reduction potential (ORP) was measured at different points along the column with a microelectrode (MI-16/800) connected to a 16-702 Flow-Thru pH reference and to the pH meter.

Microcosm experiments. Aquifer microcosms were used to investigate how exposure history and electron acceptor conditions affect substrate interactions between BTEX, EtOH, and MtBE. Microcosms were prepared with aquifer material from previously contaminated and uncontaminated sites to study biodegradation under aerobic, denitrifying, iron-reducing, sulfidogenic, and methanogenic conditions. Electron acceptors were added in excess to appropriate microcosms to avoid confounding effects associated with their depletion. Aerobic microcosms were prepared by adding 20 g of aquifer material and 80 mL of mineral medium to 125-mL amber bottles. The medium was similar to that used for the columns (Table 2), except for electron acceptor amendments. Microcosms were prepared in triplicate and capped with Mininert valves. All treatments were spiked with BTEX at about 10 mg/L total, and were fed BTEX alone, BTEX plus EtOH (100 mg/L), BTEX plus MtBE (10 mg/L) or BTEX plus EtOH and MtBE. Abiotic controls were also prepared with Kathon biocide. Oxygen was periodically added to the headspace to ensure aerobic conditions. Anaerobic microcosms were prepared similarly, except that the mineral medium was bicarbonate-buffered (625 mg/L as $CaCO_3$). All anaerobic microcosms were prepared and equilibrated inside a Coy anaerobic chamber (3-5% H_2; 5% CO_2; balance N_2). Denitrifying microcosms were fed 330 mg/L NO_3^-, iron-reducing microcosms were amended with 18.5 g/L amorphous ferric oxyhydroxide, and sulfate-reducing microcosms with 380 mg/L of SO_4^{2-}. The methanogenic microcosms contained carbonate in the basal medium (625 mg/L as $CaCO_3$) as the electron acceptor. All anaerobic microcosms were incubated under quiescent conditions at 22-28 °C inside the anaerobic chamber. BTEX, MtBE, EtOH, and methane were analyzed by gas chromatography, as described earlier. Nitrite, nitrate, sulfate, and acetate were analyzed by ion chromatography, and Fe(II) production was determined by the phenanthroline method.

RESULTS AND DISCUSSION

Effect of MtBE and EtOH on BTX Migration and Retardation. The addition of oxygenates to gasoline could affect the equilibrium partitioning relationships for BTEX compounds between aqueous, fuel and solid phases (i.e., the cosolvency effect). Specifically, MtBE or EtOH could reduce the polarity of the aqueous phase allowing higher concentrations of moderately hydrophobic compounds (e.g., BTEX) in the aqueous phase (Powers et al., 2001). In theory, the cosolvent effect exerted by either MtBE or EtOH could also enhance the mobility of dissolved BTEX compounds by decreasing sorption-related retardation. Such effects, however, are concentration-dependent, and were not observed in this work when MtBE or EtOH was fed continuously to sterile aquifer columns (as a co-contaminant with BTEX) at 10,000 mg/L (Fig. 1A). This is evident from the similarity of the breakthrough data from columns fed toluene alone or with EtOH or MtBE. Apparently, this oxygenate concentration is much lower than is required to create significant cosolvent effects (Corseuil and Fernandes, 1999). Since this EtOH concentration is unlikely to be exceeded in gasohol-contaminated sites, adding EtOH to gasoline (e.g., at 10% v/v) should not have a significant impact on BTEX retardation factors. Nevertheless, neat spills of EtOH (e.g., at a bulk terminal) could result in very high EtOH concentrations in a localized area, exerting a significant co-solvent effect that exacerbates groundwater pollution by mobilizing pre-existing petroleum product releases. This hypothesis is supported by our data (Fig. 1B). EtOH at 50% enhanced the migration of toluene, as well as benzene and o-xylene (data not shown), which traveled unretarded at same velocity as the bromide tracer.

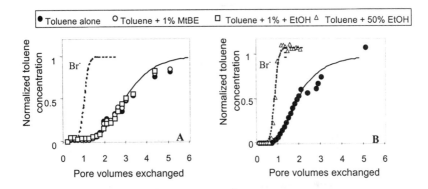

FIGURE 1. Retardation of toluene in columns fed BTX alone and with 1% EtOH or MtBE (Panel A) and with 50% EtOH (Panel B).

Effects of MtBE and EtOH on BT biodegradation and natural attenuation.
No significant decreases in benzene, toluene, EtOH and MtBE concentrations (<8%) were observed in the sterile control column, showing that volatile losses were relatively minor. All tested compounds degraded under certain conditions, except for MtBE, which was recalcitrant in all experiments.

BT were completely degraded within the first 3 cm of the inlet in columns fed BT alone or fed BT plus MtBE (Fig. 2A). MtBE did not affect the degradation of EtOH either (data not shown). On the other hand, BT degradation was adversely affected by the degradation of EtOH, which was preferentially utilized within 3 cm of the column inlet (Fig. 2B). The degradation of EtOH used up the available oxygen and exacerbated the consumption of anaerobic electron acceptors such as nitrate and sulfate (data not shown). The high electron-acceptor demand exerted by EtOH created strongly reducing conditions (-350 mV near the inlet), which are not conducive to rapid BTEX degradation (Fig. 3).

Because of the high oxygen demand exerted by gasohol spills, EtOH is likely to be degraded predominantly under anaerobic conditions. None of the products of anaerobic EtOH degradation is toxic, although some metabolites (e.g., butyrate) could have adverse aesthetic impacts. In addition, acetate and other volatile fatty acids can decrease the pH if they accumulate at high concentrations. This is illustrated in Fig. 4A, which shows acetate accumulation (1.1 g/L) only in the columns fed BT plus EtOH. Acetate accumulation caused a decrease in pH from 7.5 to as low as 6.2 (Figure 4B), even though this was a well-buffered system (1 g/L as $CaCO_3$). Poorly buffered systems could experience a greater decrease in pH that could inhibit microbial activity and the further degradation of EtOH and other compounds. Another potential concern is the accumulation of ethanol-derived methane, which could represent an explosion hazard. However, the highest methane concentration measured in the column fed BT plus EtOH was only 1.8 mg/L (Figure 4A), which is considerably below the solubility of methane (~21 mg/L at 25°C). Therefore, this experiment suggests that gasohol spills should not pose a significant explosion risk, notwithstanding the possibility that some site-specific conditions could favor more extensive methanogenesis.

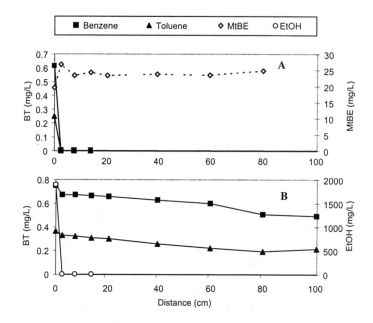

FIGURE 2. Benzene and toluene concentration profiles in columns fed MtBE (Panel A) or EtOH (Panel B) after 110 days of operation.

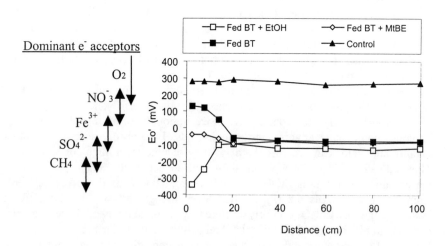

FIGURE 3. ORP profiles after 110 days of operation.

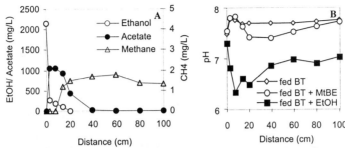

FIGURE 4. Acetate and methane concentration along the column fed BT plus EtOH (Panel A), and pH profiles (Panel B) after 226 days of operation.

Microcosms were used to determine how EtOH and MtBE affect BTEX degradation rates under different electron-acceptor conditions commonly found in contaminated aquifers experiencing natural attenuation. Response variability was also investigated by preparing microcosms with aquifer material from different sites (Table 3). BTEX and EtOH were typically degraded earlier in microcosms with previously contaminated aquifer material, although previous exposure did not always result in high degradation activity. MtBE was not degraded within 120 days under any conditions, and did not affect BTEX or EtOH degradation patterns. EtOH was typically degraded before BTEX compounds, but had a variable effect on toluene degradation as a function of electron-accepting conditions and bacterial source. In some cases, EtOH retarded toluene degradation (e.g., under methanogenic conditions for Site 1 or sulfate-reducing conditions for Site 3; Table 3), but it occasionally enhanced toluene degradation (e.g., under denitrifying conditions for Sites 3 and 4; Table 3). Enhancement of toluene degradation by EtOH may be attributable to a low initial concentration of BTEX degraders and their incidental growth during EtOH degradation.

TABLE 3: Time (days) required for 50% toluene degradation in microcosms with material from different sites under different electron acceptor conditions.

		Site*			
		1	2	3	4
Aerobic	BTEX alone	1	8	11	4
	With EtOH	1	6	14	>13
	With MtBE	1	NA	NA	7
Denitrifiying	BTEX alone	4	32	>48	51
	With EtOH	4	30	14	31
	With MtBE	4	NA	NA	36
Iron-reducing	BTEX alone	5	29	10	NA
	With EtOH	18	>70	17	NA
	With MtBE	5	NA	NA	NA
Sulfate-reducing	BTEX alone	11	>70	24	30
	With EtOH	7	>70	33	11
	With MtBE	11	NA	NA	8
Methanogenic	BTEX alone	5	57	>54	38
	With EtOH	34	35	>56	26
	With MtBE	6	NA	NA	16

*Note: Site 1 = Travis Air Force Base, BTEX and MtBE exposure history.
Site 2 = Tracy (CA), no previous BTEX exposure.
Site 3 = Northwest Terminal site, EtOH (average 160 mg/L) and BTEX exposure history.
Site 4 = Sacramento (CA), BTEX and MtBE exposure history.
NA = Data not available (i.e., not tested).

CONCLUSIONS

Cosolvent effects that increase the dissolution and migration of BTEX are unlikely to occur at the EtOH or MtBE concentrations expected at reformulated-gasoline contaminated sites (i.e., < 10,000 mg/L), but could be important when dealing with neat EtOH releases at bulk terminals.

The preferential degradation of EtOH, the accompanying depletion of O_2 and other electron acceptors, and the decrease in pH due to acetate accumulation in anaerobic zones suggest that EtOH could hinder natural attenuation of BTEX plumes. This is particularly important for the fate of benzene, which is the most toxic and the most recalcitrant of the BTEX compounds under anaerobic conditions (Heider et al. 1998). Thus, EtOH could increase the distance that benzene migrates before attenuating processes stabilizes the plume.

No effects on BTEX behavior are generally expected by the presence of MtBE. Nevertheless, MtBE itself is a major concern in drinking water supplies because of consumer complaints of the pungent odor and potential health effects. Therefore, the impacts that result with the use of EtOH seems to be less significant and more manageable than those associated with using MtBE as a fuel oxygenate.

ACKNOWLEDGMENTS

This work was funded by API, the State of California through Lawrence Livermore National Laboratory, and CAPES - Brazil. We thank Jeff Pullen and Craig Just for their technical assistance, and T. Buscheck (Chevron), W. Day (Travis AFB), G. Mongano (US Bureau of Reclamation), and M. Peterson (ETIC Engineering) for donating aquifer materials for microcosm studies.

REFERENCES

Alvarez P.J.J., L.C. Cronkhite, and C.S. Hunt. 1998. Use of benzoate to establish reactive buffer zones for enhanced attenuation of BTX migration. *Environ. Sci. Technol.* 32(4): 509-515.

Corseuil H.X. and M. Fernandes. 1999. Cosolvency effect in aquifers contaminated with ethanol-amended gasoline. In: Alleman B.C. and A. Leeson (eds.) *Natural attenuation of chlorinated petroleum hydrocarbons, and other organic compounds*. Fifth International In situ and On-site Bioremediation Symposium., Battelle Press. 5(1): 135-140.

Federal Register. 2000. Methyl tertiary butyl ether (MTBE): Advance notice of intent to initiate rulemaking under the toxic substance control act to eliminate or limit the use of MTBE as a fuel additive in gasoline. 65(58), 16094. U.S. Government Printing Office, Washington, D.C.

Heider J., A.M. Spormann, H.R. Beller, and F. Widdel. 1998. Anaerobic bacterial metabolism of hydrocarbons. *FEMS Microbiology Reviews* 22:459-473.

Powers S.E., C.S. Hunt, S.E. Heermann, H.X. Corseuil, D. Rice, and P.J.J Alvarez. 2001. The transport and fate of ethanol and BTEX in groundwater contaminated by gasohol. *Crit. Rev. Environ. Sci. Technol.* 31(1):79-123.

van Gunten and Jurg Zobrist. 1993. Biogeochemical changes in groundwater-infiltration systems: Columns studies. *Geochim. Cosmochim. Acta* 57: 385-3906.

ETHANOL IN GROUNDWATER AT A NORTHWEST TERMINAL

Timothy E. Buscheck and Kirk O'Reilly (Chevron Research and Technology Company, Richmond, CA, USA)
Gerard Koschal (PNG Environmental, Tigard, OR, USA)
Gerald O'Regan, (Chevron Products Company, San Ramon, CA, USA)

ABSTRACT: In March 1999, a 19,000-gallon release of neat ethanol occurred from an aboveground storage tank at a Bulk Fuel Terminal located in the Northwest United States ("Northwest Terminal"). Ethanol is completely miscible in water and at high concentrations (>20,000 mg/L) can enhance the solubilization of benzene, toluene, ethylbenzene, and the xylenes (BTEX) from nonaqueous phase liquid (NAPL). Ethanol can be degraded in both aerobic and anaerobic environments. The presence of ethanol appears to enhance the mobility of NAPL, as shown by increasing apparent NAPL thickness in two monitoring wells. Cosolvent effects of ethanol and lower benzene transformation rates are suggested by increasing benzene concentrations in one downgradient monitoring well. The presence of ethanol has created a strongly anaerobic groundwater system, demonstrated by low or nondetectable dissolved oxygen, depleted sulfate and nitrate, and elevated methane concentrations. The monitoring well network is adequate to delineate the ethanol plume.

INTRODUCTION

The Northwest Terminal began petroleum-processing operations in 1911, distributing and blending a variety of refined petroleum products, including gasoline, diesel, Bunker C oil, stove oil, turbine oil, transmission fluid and lubrication oils (Kline, 2001). Two other terminals located to the southeast and northwest border the Northwest Terminal.

On March 20, 1999, a 19,000-gallon release of neat ethanol occurred from Tank 58 at the Northwest Terminal. Historical groundwater monitoring data were available from existing monitoring wells to delineate a pre-existing dissolved hydrocarbon plume. On March 30, 1999 four existing monitoring wells were sampled and analyzed for BTEX, total petroleum hydrocarbons (TPH), and ethanol. In June and July of 1999, nine monitoring wells were installed to delineate ethanol and hydrocarbons in the subsurface. These wells were sampled six times between June 1999 and December 2000. Since May 2000 the groundwater sampling protocol has included analytes for in situ bioremediation. In December 2000, eight additional monitoring wells were installed.

GEOLOGIC AND HYDROGEOLOGIC SETTING

The hydrogeologic setting consists of a layer of fill on top of alluvium (Kline, 2001). The fill is very loose to medium dense and fine to medium grained sand and silty sand. The thickness of the fill material ranges from nonexistent to greater than 30 feet (9 m). In general, the alluvium is very soft to medium stiff clayey silt with sand and organics. The alluvium is often interbedded with silty clays and clays. The alluvium occurs to a depth of approximately 50 feet (15.2 m) below grade, where basaltic material is present.

The fill and the alluvium are hydraulically connected. The units discharge to a local river, located approximately 1500 feet to the east. The fill may be locally perched on the alluvial unit. The direction of groundwater flow is generally toward the river at approximately 0.01 ft/ft. The fill layer is the primary zone for the occurrence of hydrocarbons. The underlying alluvium is, in some locations, less permeable and may provide an aquitard.

Water levels beneath the site are influenced by annual precipitation cycles and by river stage fluctuations. The depths to water typically range from 4 to 14 feet (1.2 to 4.3 m) below grade. Based on an effective porosity of 40 percent and a hydraulic conductivity of 35 feet per day (10.7 m per day) (based on field tests), the groundwater velocity within the fill material is estimated to be 300 to 400 feet per year (92 to 123 m per year) (Kline, 2001).

ETHANOL DEGRADATION PATHWAYS

In the presence of ethanol, biodegradation of BTEX in groundwater may be inhibited, potentially increasing hydrocarbon plume lengths. Ethanol can be degraded in both aerobic and anaerobic environments. Microbial degradation of ethanol generates a variety of metabolic intermediates and end products (Alvarez and Hunt, 1999). Oxygen can be quickly depleted by microbial aerobic respiration in hydrocarbon-contaminated aquifers. Therefore, ethanol is likely to be degraded under anaerobic conditions at field sites.

Aerobic Degradation. Most common aerobic bacteria can oxidize ethanol through the Krebs cycle (Brock and Madigan, 1991). The intermediates include acetaldehyde and acetyl coenzyme A (acetyl-CoA). Carbon dioxide (CO_2) is a final degradation product. The intermediates of the common metabolic pathways are not toxic. They are metabolized rapidly intracellularly and are unlikely to accumulate in groundwater; one exception is acetic acid bacteria, which excrete acetate (Alvarez and Hunt, 1999). However, acetate will degrade under either aerobic or anaerobic conditions.

Anaerobic Pathways. Microorganisms that can ferment ethanol are ubiquitous (Alvarez and Hunt, 1999). Ethanol is a common intermediate in the anaerobic food chain, where labile organic matter is degraded to nontoxic products by the combined action of several different types of bacteria. These nontoxic products include acetate, CO_2, methane (CH_4), and hydrogen gas (H_2) (Alvarez and Hunt, 1999). First stage fermenters produce simple organic acids, alcohols, hydrogen gas, and CO_2. In the second stage, obligate proton reducers, sulfate-reducers, and acetogens produce acetate, H_2, and CO_2. In the third stage, methanogens mineralize acetate to CO_2 and CH_4 (Alvarez and Hunt, 1999). The most important CO_2-reducing bacteria are the methanogens (Brock and Madigan, 1991).

Ethanol Degradation Rates in Groundwater. The rate of ethanol degradation under either aerobic or anaerobic conditions is faster than the degradation rates of other gasoline constituents (Corseuil et al., 1998). First-order, laboratory ethanol degradation half-lives measured by Corseuil et al. (1998) varied from 1 to 7 days, depending on the electron acceptor. The only field-scale studies with fuel-alcohol releases have been conducted with methanol. In a field study at Canadian Forces Base Borden, Ontario, Canada, a blend of 85% methanol and 15% gasoline (M85 fuel) was released in a shallow, sandy aquifer (API, 1994). Methanol was biodegraded from an initial concentration of 7000 mg/L to below 1 mg/L in 476 days, equating to a degradation rate of 0.019 day^{-1} (36-day half-life). The biodegradation rates for methanol and ethanol are expected to be fairly similar.

Potential Effects of Ethanol on BTEX Biodegradation. Ethanol concentrations above 40,000 mg/L have been shown to inhibit oxygen utilization by indigenous organisms. (Hunt et al., 1997). With 10% ethanol in ethanol-blended gasoline, ethanol concentrations measured at underground storage tank (UST) release sites should not have a significant impact on BTEX biodegradation (da Silva et al., 2001). However, depletion of electron acceptors by ethanol degrading organisms will reduce their availability to BTEX degraders. Competition for electron acceptors could lead to a decrease in the BTEX intrinsic biodegradation rate and potentially result in longer BTEX plumes.

In microcosm experiments conducted by Hunt et al. (1997), the presence of 300 mg/L or less of ethanol did not slow benzene and toluene degradation by *Pseudomonas putida* F1 (PpF1), but a slight inhibitory effect was observed at 500 mg/L ethanol. However, the simultaneous utilization of ethanol and toluene as growth substrates for PpF1 represents a caveat against generalizations about the effect of fuel additives on BTEX degradation patterns (Hunt et al., 1997). Additional research is needed to define the effect of ethanol on BTEX biodegradation.

Corseuil et al. (1998) conducted laboratory experiments under aerobic conditions. They found that ethanol depleted the available oxygen, creating anaerobic conditions. These authors suggest that because benzene degrades more slowly under anaerobic conditions, it could migrate further in groundwater than it would without ethanol's presence.

EVIDENCE FOR BIODEGRADATION OF ETHANOL

In Figure 1 the ethanol concentrations are posted for June 1999 and December 2000. The variability in ethanol's practical quantitation limit (PQL) made it difficult to contour. June 1999 ethanol concentrations were not available for three of the wells shown in Figure 1 (CR-7, 8, and 10); February 2000 results are posted for those wells because the best detection limits were available on that date. The direction of groundwater flow and former Tank 58, the location of the ethanol release, are also shown in Figure 1. For the monitoring dates shown in Figure 1, ethanol was detected in only four wells, CR-12, 13, 14, and 19. These wells are located within approximately 100 feet of Tank 58. Between June 1999 and December 2000, significant reduction in ethanol concentration is observed in each of these wells.

Evidence of Rapid Plume Attenuation. During the 18-month monitoring period (June 1999 through December 2000) the furthest downgradient occurrence of ethanol was in CR-16 at 4,170 ug/L on September 20, 1999. Six months after the release, ethanol had migrated 250 feet (76 m) from Tank 58, consistent with groundwater velocities. This was the only date on which ethanol was detected in CR-16 and ethanol has never been detected above the PQL in any of the wells downgradient of CR-16, suggesting rapid attenuation terminated plume expansion between 6 and 12 months after the release.

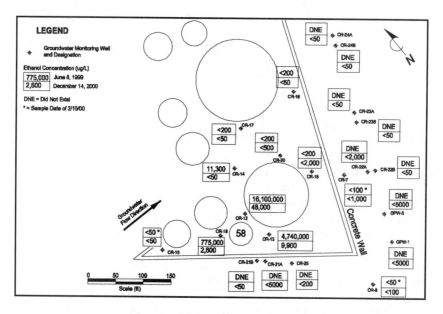

FIGURE 1. June 1999 and December 2000 Ethanol Concentrations

In Situ Bioremediation Sampling Protocol. In May 2000 the groundwater sampling protocol was expanded to include analytes for in situ bioremediation. Those analytes include dissolved oxygen (DO), methane, nitrate, sulfate, CO_2, acetate, and dC^{13} (carbon-13 isotope for CO_2). These parameters were measured in May, August, and December 2000. Methane concentrations ranging from 2,000 to 18,000 ug/L, low or nondetectable DO, and depleted sulfate and nitrate demonstrate strongly reducing conditions. The spatial distribution of methane is shown in Figure 2. Acetate, an intermediate product of ethanol biodegradation, was measured in downgradient monitoring wells in December 2000.

December 2000 Groundwater Investigation. In December 2000, eight new monitoring wells were installed to define the vertical hydraulic gradient and further delineate the vertical and lateral extent of ethanol. Nested wells were screened to monitor the shallow water table (A) and the deeper alluvium (B) at a depth of 35 to 40 feet (10.7 to 12.2 m). Within 50 feet (15.2 m) south of Tank 58,

CR-21A and CR-21B show a downward gradient. A and B well pairs completed 250 to 350 feet (76 to 107 m) downgradient of former Tank 58 (CR-22, 23, and 24 well pairs), each show an upward hydraulic gradient from the alluvium to the fill material. GPW-1 and GPW-3 are shallow water table wells that were included in the ethanol investigation for the first time in December 2000. Ethanol was not detected above the PQL in any of the downgradient monitoring wells sampled in December 2000 for the first time (Figure 1).

Figure 2 illustrates the December 2000 methane concentration contours. The highest methane concentrations occur in the footprint of the ethanol plume. December 2000 methane concentrations vary from less than 1,000 ug/L to almost 18,000 ug/L. These methane concentrations indicate the presence of ethanol in the subsurface has depleted electron acceptors, creating strongly reducing, methanogenic conditions. Based on the December 2000 groundwater results, the ethanol plume appears to be adequately defined by the existing monitoring well network.

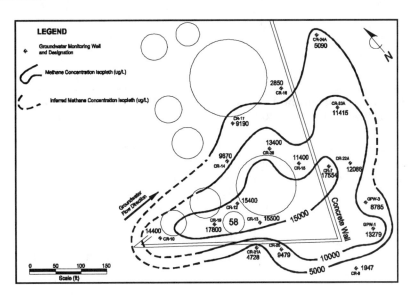

FIGURE 2. December 2000 Methane Concentration Contours

Evidence of Residual Ethanol in Vadose Zone Soils. Monitoring wells CR-12 and CR-13 are located downgradient, within 40 feet of former Tank 58 and within the existing ethanol plume. The highest ethanol concentrations ever measured at the terminal occurred in CR-12 and CR-13, at 16,100,000 ug/L and 4,740,000 ug/L, respectively. These concentrations were measured less than three months after the March 1999 ethanol release (June 9, 1999). Figure 3 is a plot for CR-12 ethanol concentration and depth to groundwater versus time. Between June 1999 and August 2000, ethanol concentrations followed a steady decline in CR-12, further evidence of ethanol attenuation. On August 15, 2000 ethanol was not detected above the PQL of 50 ug/L. On December 15, 2000 the ethanol concentration was 48,000 ug/L (PQL of 2000 ug/L). Total petroleum hydrocarbon concentration increased from 6800 to 16,000 ug/L between August and December, which likely contributed to the higher PQL in December. The duplicate groundwater samples obtained in both August and December 2000 show good agreement in the ethanol concentrations. Figure 3 suggests the depth to groundwater did not vary between August and December. However, significant rainfall in early December, prior to the sampling event, likely explains the increased ethanol concentrations in CR-12 resulting from infiltrating rainwater contacting residual ethanol that remained in the vadose zone.

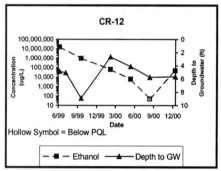

FIGURE 3. Ethanol and Depth to Groundwater in Well CR-12

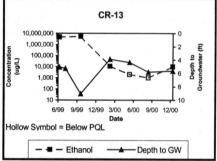

FIGURE 4. Ethanol and Depth to Groundwater in Well CR-13

Figure 4 is a plot for CR-13 ethanol concentration and depth to groundwater versus time. In August 2000 ethanol was not detected at the PQL of 1000 ug/L, almost 5000 times below the June 1999 concentration. In December ethanol was detected at 9,900 ug/L. As was observed in CR-12, the rebound in ethanol concentration suggests that residual ethanol remains in the vadose zone.

IMPACT OF ETHANOL ON HYDROCARBONS

The presence of ethanol has impacted hydrocarbon in both the NAPL and dissolved phases.

Evidence of Increased Mobility of Pre-existing NAPL. Monitoring well CR-19 is located within 20 feet (9 m), southwest of former Tank 58. The greatest NAPL thickness measured in the vicinity of the ethanol release has occurred in well CR-19. Figure 5 illustrates NAPL thickness and groundwater elevation for CR-19. These plots show the water table decline of 2 feet (0.6 m) through the summer of 1999 and the water table rise by as much as 3 feet (0.9 m) in the winter of 2000. Through the winter of 2000, NAPL thickness remained constant in CR-19 (approximately 2.5 feet (0.8 m)). In May 2000 NAPL was bailed from this well and recovered very slowly. However, the NAPL thickness of 2.79 feet (0.85 m) in December 2000 is the greatest thickness measured over the 18-month monitoring period. NAPL was also present in well CR-15 (downgradient of Tank 58) and varied from nonmeasurable in June 1999 to 0.44 feet (0.13 m) in

September 1999. This NAPL occurrence may also have been influenced by the presence of ethanol, detected on one sampling event in CR-15 above the PQL (4,000 ug/L in February 2000).

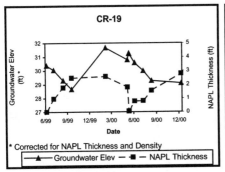
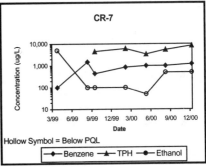

FIGURE 5. NAPL and Groundwater Elevation in Well CR-19

FIGURE 6. Ethanol, Benzene, and TPH in Well CR-7

Evidence of Impact on Dissolved Phase Benzene. Well CR-7 is located approximately 200 feet (61 m) downgradient of former Tank 58 (within 50 feet of CR-15). Ethanol has never been detected in CR-7 above the PQL. Figure 6 is a plot of ethanol, benzene, and TPH-gasoline versus time in CR-7. The March 30, 1999 sampling event occurred ten days after the ethanol release and given the distance between Tank 58 and CR-7, those analytical results provide a baseline condition. Benzene concentration increased by a factor of 15 in this well between March 30 and August 17, 1999. Throughout 2000, benzene concentrations remained about ten times above the March 1999 levels. Two factors could contribute to increased benzene concentrations in CR-7: (1) enhanced dissolution resulting from NAPL occurrence in upgradient CR-15 and (2) upgradient, elevated ethanol concentrations in CR-12 and CR-13, exerting a high demand for electron acceptors. CR-7 provides potential evidence for (1) cosolvent effects of ethanol on benzene and (2) lower benzene transformation rates in the presence of ethanol.

SUMMARY

Seventeen new monitoring wells were installed over an 18-month period in response to a March 1999, 19,000-gallon neat ethanol release. These wells provide adequate delineation of the ethanol plume. Within six months after the release, ethanol was detected 250 feet (76 m) downgradient (CR-16), but shortly afterward the plume appeared to attenuate. Ethanol concentrations have significantly declined in two monitoring wells directly downgradient of the ethanol release (CR-12 and CR-13). December 2000 ethanol concentrations increased in these two wells, compared to earlier trends, suggesting some residual ethanol remains in the vadose zone.

The presence of ethanol has affected pre-existing petroleum hydrocarbon in both the NAPL and the dissolved phases. However, ethanol concentrations at

UST release sites (10% ethanol in ethanol-gasoline mixtures) are not likely to have the same impact on a hydrocarbon plume. NAPL thickness in one near-source monitoring well (CR-19) has exceeded 2 feet (0.6 m) and is sustained during water table fluctuations. March 1999 data (pre-release) from the downgradient well, CR-7, coupled with increasing benzene concentrations in that well, suggest a possible cosolvent effect and lower benzene transformation rates resulting from the presence of ethanol.

The presence of ethanol has created a strongly anaerobic groundwater system, demonstrated by low or nondetectable DO and depleted sulfate and nitrate. There is evidence for ethanol biodegradation under methanogenic conditions, demonstrated by elevated methane concentrations and declining ethanol concentrations throughout the plume.

ACKNOWLEDGEMENTS

Chevron Products Company funded the groundwater investigation. We thank Chuck Zuspan of Chevron Research and Technology Company for his assistance in the data evaluation. We also thank Paul Ecker of PNG Environmental for conducting fieldwork in this investigation and Harry Beller of Lawrence Livermore National Laboratory for providing acetate analyses.

REFERENCES

Alvarez, P.J.J. and C.S. Hunt. 1999. "Volume 4: Potential Ground and Surface Water Impacts, Chapter 3: The Effect of Ethanol on BTEX Biodegradation and Natural Attenuation." G. Cannon and D. Rice, (Eds.) Published by the State of California, UCRL-AR-135949. (http://www-erd.llnl.gov/ethanol/)

API. 1994. *Transport and Fate of Dissolved Methanol, Methyl-Tertiary-Butyl-Ether, and Monoaromatic Hydrocarbons in a Shallow Aquifer.* API Publication Number 4601, April 1994.

Brock, T.D. and M.T. Madigan. 1991. *Biology of Microorganisms.* Prentice Hall, Englewood Cliffs, N.J.

Corseuil, H.X. and P.J.J. Alvarez. 1996. "Natural Bioremediation Perspective for BTX-Contaminated Groundwater in Brazil: Effect of Ethanol." *Water Science Technology* 34(7-8): 311-318.

Corseuil, H.X., C.S. Hunt, R. dos Santos Ferreira, and P.J.J. Alvarez. 1998. "The Influence of the Gasoline Oxygenate Ethanol on Aerobic and Anaerobic BTX Biodegradation." *Water Research* 32(7): 2065-2072.

da Silva, M.L.B, G.M Ruiz, J.M. Fernandez, H.R. Beller, and P.J.J. Alvarez. 2001. "Effect of Ethanol versus MTBE on BTEX Natural Attenuation." In Situ and On-Site Bioremediation: The Sixth International Symposium. San Diego, CA, June 4-7, 2001.

Hunt, C.S., P.J.J. Alvarez, R. dos Santos Ferreira, and H.X. Corseuil. 1997. "Effect of Ethanol on Aerobic BTEX Degradation." In B.C. Alleman and A.L. Leeson (Eds.) *In Situ and On-Site Bioremediation*, 4(1): 49-54. Battelle Press, Columbus, OH.

Kline, K. 2001. Personal Communication. KHM Environmental Management, Inc.

2001 AUTHOR INDEX

This index contains names, affiliations, and volume/page citations for all authors who contributed to the ten-volume proceedings of the Sixth International In Situ and On-Site Bioremediation Symposium (San Diego, California, June 4-7, 2001). Ordering information is provided on the back cover of this book. The citations reference the ten volumes as follows:

6(1): Magar, V.S., J.T. Gibbs, K.T. O'Reilly, M.R. Hyman, and A. Leeson (Eds.), *Bioremediation of MTBE, Alcohols, and Ethers*. Battelle Press, Columbus, OH, 2001. 249 pp.

6(2): Leeson, A., M.E. Kelley, H.S. Rifai, and V.S. Magar (Eds.), *Natural Attenuation of Environmental Contaminants*. Battelle Press, Columbus, OH, 2001. 307 pp.

6(3): Magar, V.S., G. Johnson, S.K. Ong, and A. Leeson (Eds.), *Bioremediation of Energetics, Phenolics, and Polycyclic Aromatic Hydrocarbons*. Battelle Press, Columbus, OH, 2001. 313 pp.

6(4): Magar, V.S., T.M. Vogel, C.M. Aelion, and A. Leeson (Eds.), *Innovative Methods in Support of Bioremediation*. Battelle Press, Columbus, OH, 2001. 197 pp.

6(5): Leeson, A., E.A. Foote, M.K. Banks, and V.S. Magar (Eds.), *Phytoremediation, Wetlands, and Sediments*. Battelle Press, Columbus, OH, 2001. 383 pp.

6(6): Magar, V.S., F.M. von Fahnestock, and A. Leeson (Eds.), *Ex Situ Biological Treatment Technologies*. Battelle Press, Columbus, OH, 2001. 423 pp.

6(7): Magar, V.S., D.E. Fennell, J.J. Morse, B.C. Alleman, and A. Leeson (Eds.), *Anaerobic Degradation of Chlorinated Solvents*. Battelle Press, Columbus, OH, 2001. 387 pp.

6(8): Leeson, A., B.C. Alleman, P.J. Alvarez, and V.S. Magar (Eds.), *Bioaugmentation, Biobarriers, and Biogeochemistry*. Battelle Press, Columbus, OH, 2001. 255 pp.

6(9): Leeson, A., B.M. Peyton, J.L. Means, and V.S. Magar (Eds.), *Bioremediation of Inorganic Compounds*. Battelle Press, Columbus, OH, 2001. 377 pp.

6(10): Leeson, A., P.C. Johnson, R.E. Hinchee, L. Semprini, and V.S. Magar (Eds.), *In Situ Aeration and Aerobic Remediation*. Battelle Press, Columbus, OH, 2001. 391 pp.

Aagaard, Per (University of Oslo/NORWAY) 6(2):181
Aarnink, Pedro J.P. (Tauw BV/THE NETHERLANDS) 6(10):253
Abbott, James E. (Battelle/USA) 6(5):231, 237
Accashian, John V. (Camp Dresser & McKee, Inc./USA) 6(7):133
Adams, Daniel J. (Camp Dresser & McKee, Inc./USA) 6(8):53
Adams, Jack (Applied Biosciences Corporation/USA) 6(9):331
Adriaens, Peter (University of Michigan/USA) 6(8):19, 193
Adrian, Neal R. (U.S. Army Corps of Engineers/USA) 6(6):133
Agrawal, Abinash (Wright State University/USA) 6(5):95
Aiken, Brian S. (Parsons Engineering Science/USA) 6(2): 65, 189

Aitchison, Eric (Ecolotree, Inc./USA) 6(5):121
Al-Awadhi, Nader (Kuwait Institute for Scientific Research/KUWAIT) 6(6):249
Alblas, B. (Logisticon Water Treatment/THE NETHERLANDS) 6(8):11
Albores, A. (CINVESTAV-IPN/MEXICO) 6(6):219
Al-Daher, Reyad (Kuwait Institute for Scientific Research/KUWAIT) 6(6):249
Al-Fayyomi, Ihsan A. (Metcalf & Eddy, Inc./USA) 6(7):173
Al-Hakak, A. (McGill University/CANADA) 6(9):139
Allen, Harry L. (U.S. EPA/USA) 6(3):259
Allen, Jeffrey (University of Cincinnati/USA) 6(9):9
Allen, Mark H. (Dames & Moore/USA) 6(10):95
Allende, J.L. (Universidad Complutense/SPAIN) 6(4):29
Alonso, R. (Universidad Politecnica/SPAIN) 6(6):377
Alphenaar, Arne (TAUW bv/THE NETHERLANDS) 6(7):297
Alvarez, Pedro J. J. (University of Iowa/USA) 6(1):195; 6(3):1; 6(8):147, 175
Alvestad, Kimberly R. (Earth Tech/USA) 6(3):17
Ambert, Jack (Battelle Europe/SWITZERLAND) 6(6):241
Amezcua-Vega, Claudia (CINVESTAV-IPN/MEXICO) 6(3):243
Amy, Penny (University of Nevada Las Vegas/USA) 6(9):257
Andersen, Peter F. (GeoTrans, Inc./USA) 6(10):163
Anderson, Bruce (Plan Real AG/AUSTRALIA) 6(2):223
Anderson, Jack W. (RMT, Inc./USA) 6(10):201
Anderson, Todd (Texas Tech University/USA) 6(9):273
Andreotti, Giorgio (ENI Sop.A.) 6(5):41

Andretta, Massimo (Centro Ricerche Ambientali Montecatini/ITALY) 6(4):131
Andrews, Eric (Environmental Management, Inc./USA) 6(10):23
Andrews, John (SHN Consulting Engineers & Geologists, Inc./USA) 6(3):83
Archibald, Brent B. (Exxon Mobil Environmental Remediation/USA) 6(8):87
Archibold, Errol (Spelman College/USA) 6(9):53
Aresta, Michele (Universita di Catania/ITALY) 6(3):149
Arias, Marianela (PDVSA Intevep/VENEZUELA) 6(6):257
Atagana, Harrison I. (Mangosuthu Technikon/REP OF SOUTH AFRICA) 6(6):101
Atta, Amena (U.S. Air Force/USA) 6(2):73
Ausma, Sandra (University of Guelph/CANADA) 6(6):185
Autenrieth, Robin L. (Texas A&M University/USA) 6(5): 17, 25
Aziz, Carol E. (Groundwater Services, Inc./USA) 6(7):19; 6(8):73
Azizian, Mohammad (Oregon State University/USA) 6(10): 145, 155

Babel, Wolfgang (UFZ Center for Environmental Research/GERMANY) 6(4):81
Bae, Bumhan (Kyungwon University/REPUBLIC OF KOREA) 6(6):51
Baek, Seung S. (Kyonggi University/REPUBLIC OF KOREA) 6(1):161
Bagchi, Rajesh (University of Cincinnati/USA) 6(5):243, 253, 261
Baiden, Laurin (Clemson University/USA) 6(7):109
Bakker, C. (IWACO/THE NETHERLANDS) 6(7):141
Balasoiu, Cristina (École Polytechnique de Montreal/CANADA) 6(9):129
Balba, M. Talaat (Conestoga-Rovers & Associates/USA) 6(1):99; 6(6):249; 6(10):131

Author Index

Banerjee, Pinaki (Harza Engineering Company, Inc./USA) 6(7):157
Bankston, Jamie L. (Camp Dresser and McKee Inc./USA) 6(5):33
Barbé, Pascal (Centre National de Recherche sur les Sites et Sols Pollués/FRANCE) 6(2):129
Barcelona, Michael J. (University of Michigan/USA) 6(8):19, 193
Barczewski, Baldur (Universitat Stuttgart/GERMANY) 6(2):137
Barker, James F. (University of Waterloo/CANADA) 6(8):95
Barnes, Paul W. (Earth Tech, Inc./USA) 6(3): 17, 25
Basel, Michael D. (Montgomery Watson Harza/USA) 6(10):41
Baskunov, Boris B. (Russian Academy of Sciences/RUSSIA) 6(3):75
Bastiaens, Leen (VITO/BELGIUM) 6(4):35; 6(9):87
Batista, Jacimaria (University of Nevada Las Vegas/USA) 6(9): 257, 265
Bautista-Margulis, Raul G. (Centro de Investigacion en Materiales Avanzados/MEXICO) 6(6):361
Becker, Paul W. (Exxon Mobil Refining & Supply/USA) 6(8):87
Beckett, Ronald (Monash University/AUSTRALIA) 6(4):1
Beckwith, Walt (Solutions Industrial & Environmental Services/USA) 6(7):249
Beguin, Pierre (Institut Pasteur/FRANCE) 6(1):153
Behera, N. (Sambalpur University/INDIA) 6(9):173
Bell, Nigel (Imperial College London/UK) 6(10):123
Bell, Mike (Coats North America/USA) 6(7):213
Beller, Harry R. (Lawrence Livermore National Laboratory/USA) 6(1):195
Belloso, Claudio (Facultad Catolica de Quimica e Ingenieria/ARGENTINA) 6(6): 235, 303
Benner, S. G. (Stanford University/USA) 6(9):71
Bensch, Jeffrey C. (GeoTrans, Inc/USA) 6(7):221

Béron, Patrick (Université du Québec à Montréal/CANADA) 6(3):165
Berry, Duane F. (Virginia Polytechnic Institute & State University/USA) 6(2):105
Betts, W. Bernard (Cell Analysis Ltd./UK) 6(6):27
Billings, Bradford G. (Brad) (Billings & Associates, Inc./USA) 6(1):115
Bingler, Linda (Battelle Sequim/USA) 6(5):231, 237
Birkle, M. (Fraunhofer Institute/GERMANY) 6(2):137
Bitter, Paul (URS Corporation./USA) 6(2):261
Bittoni, A. (EniTecnologie/ITALY) 6(6):173
Bjerg, Poul L (Technical University of Denmark/DENMARK) 6(2):11
Blanchet, Denis (Institut Français du Pétrole/FRANCE) 6(3):227
Bleckmann, Charles A. (Air Force Institute of Technology/USA) 6(2):173
Blokzijl, R. (DHV Environment and Infrastructure/THE NETHERLANDS) 6(8):11
Blowes, David (University of Waterloo/CANADA) 6(9):71
Bluestone, Simon (Montgomery Watson/ITALY) 6(10):41
Boben, Carolyn (Williams/USA) 6(1):175
Böckle, Karin (Technologiezentrum Wasser/GERMANY) 6(8):105
Boender, H. (Logisticon Water Treatment/THE NETHERLANDS) 6(8):11
Böhler, Anja (BioPlanta GmbH/GERMANY) 6(3):67
Bonner, James S. (Texas A&M University/USA) 6(5):17, 25
Bononi, Vera Lucia Ramos (Instituto de Botânica/BRAZIL) 6(3):99
Bonsack, Laurence T. (Aerojet/USA) 6(9):297
Borazjani, Abdolhamid (Mississippi State University/USA) 6(5):329; 6(6):279

Borden, Robert C. (Solutions Industrial & Environmental Services/USA) 6(7):249
Bornholm, Jon (U.S. EPA/USA) 6(6):81
Bosco, Francesca (Politecnico di Torino/ITALY) 6(3):211
Bosma, Tom N.P. (TNO Environment/THE NETHERLANDS) 6(7):61
Bourquin, Al W. (Camp Dresser & McKee Inc./USA) 6(5):33; 6(6):81; 6(7):133,
Bouwer, Edward J. (Johns Hopkins University/USA) 6(2):19
Bowman, Robert S. (New Mexico Institute of Mining & Technology/USA) 6(8):131
Boyd, Sian (CEFAS Laboratory/UK) 6(10):337
Boyd-Kaygi, Patricia (Harding ESE/USA) 6(10):231
Boyle, Susan L. (Haley & Aldrich, Inc./USA) 6(7):27, 281
Brady, Warren D. (IT Corporation/USA) 6(9):215
Breedveld, Gijs (University of Oslo/NORWAY) 6(2):181
Bregante, M. (Istituto di Cibernetica e Biofisica/ITALY) 6(5):157
Brenner, Richard C. (U.S. EPA/USA) 6(5):231, 237
Breteler, Hans (Oostwaardhoeve Co./THE NETHERLANDS) 6(6):59
Bricka, Mark R. (U.S. Army Corps of Engineers/USA) 6(9):241
Brickell, James L. (Earth Tech, Inc./USA) 6(10):65
Brigmon, Robin L. (Westinghouse Savannah River Co/USA) 6(7):109
Britto, Ronnie (EnSafe, Inc./USA) 6(9):315
Brossmer, Christoph (Degussa Corporation/USA) 6(10):73
Brown, Bill (Dunham Environmental Services/USA) 6(6):35
Brown, Kandi L. (IT Corporation/USA) 6(1):51
Brown, Richard A. (ERM, Inc./USA) 6(7):45, 213
Brown, Stephen (Queen's University/CANADA) 6(2):121

Brown, Susan (National Water Research Institute/CANADA) 6(7):321, 333, 341
Brubaker, Gaylen (ThermoRetec North Carolina Corp./USA) 6(7):1
Bruce, Cristin (Arizona State University/USA) 6(8):61
Bruce, Neil C. (University of Cambridge/UK) 6(5):69
Buchanan, Gregory (Tait Environmental Management, Inc./USA) 6(10):267
Bucke, Christopher (University of Westminster/UK) 6(3):75
Bulloch, Gordon (BAE Systems Properties Ltd./UK) 6(6):119
Burckle, John (U.S. EPA/USA) 6(9):9
Burden, David S. (U.S. EPA/USA) 6(2):163
Burdick, Jeffrey S. (ARCADIS Geraghty & Mills/USA) 6(7):53
Burgos, William (The Pennsylvania State University/USA) 6(8):201
Burken, Joel G. (University of Missouri-Rolla/USA) 6(5):113, 199
Burkett, Sharon E. (ENVIRON International Corp./USA) 6(7):189
Burnell, Daniel K. (GeoTrans, Inc./USA) 6(2):163
Burns, David A. (ERM, Inc./USA) 6(7):213
Burton, Christy D. (Battelle/USA) 6(1):137; 6(10):193
Buscheck, Timothy E. (Chevron Research & Technology Co/USA) 6(1): 35, 203
Buss, James A. (RMT, Inc./USA) 6(2):97
Butler, Adrian P. (Imperial College London/UK) 6(10):123
Butler, Jenny (Battelle/USA) 6(7):13
Büyüksönmez, Fatih (San Diego State University/USA) 6(10):301

Caccavo, Frank (Whitworth College/USA) 6(8):1
Callender, James S. (Rockwell Automation/USA) 6(7):133
Calva-Calva, G. (CINVESTAV-IPN/MEXICO) 6(6):219
Camper, Anne K. (Montana State University/USA) 6(7):117

Author Index

Camrud, Doug (Terracon/USA) 6(10):15
Canty, Marietta C. (MSE Technology Applications/USA) 6(9):35
Carman, Kevin R. (Louisiana State University/USA) 6(5):305
Carrera, Paolo (Ambiente S.p.A./ITALY) 6(6):227
Carson, David A. (U.S. EPA/USA) 6(2):247
Carvalho, Cristina (Clemson University/USA) 6(7):109
Case, Nichole L. (Haley & Aldrich, Inc./USA) 6(7):27, 281
Castelli, Francesco (Universita di Catania/ITALY) 6(3):149
Cha, Daniel K. (University of Delaware/USA) 6(6):149
Chaney, Rufus L. (U.S. Department of Agriculture/USA) 6(5):77
Chang, Ching-Chia (National Chung Hsing University/TAIWAN) 6(10):217
Chang, Soon-Woong (Kyonggi University/REPUBLIC OF KOREA) 6(1):161
Chang, Wook (University of Maryland/USA) 6(3):205
Chapuis, R. P. (École Polytechnique de Montréal/CANADA) 6(4):139
Charrois, Jeffrey W.A. (Komex International, Ltd./CANADA) 6(4):7
Chatham, James (BP Exploration/USA) 6(2):261
Chekol, Tesema (University of Maryland/USA) 6(5):77
Chen, Abraham S.C. (Battelle/USA) 6(10):245
Chen, Chi-Ruey (Florida International University/USA) 6(10):187
Chen, Zhu (The University of New Mexico/USA) 6(9):155
Cherry, Jonathan C. (Kennecott Utah Copper Corp/USA) 6(9):323
Child, Peter (Investigative Science Inc./CANADA) 6(2):27
Chino, Hiroyuki (Obayashi Corporation/JAPAN) 6(6):249
Chirnside, Anastasia E.M. (University of Delaware/USA) 6(6):9
Chiu, Pei C. (University of Delaware/USA) 6(6):149
Cho, Kyung-Suk (Ewha University/REPUBLIC OF KOREA) 6(6):51
Choung, Youn-kyoo (Yonsei University/REPUBLIC OF KOREA) 6(6):51
Clement, Bernard (École Polytechnique de Montréal/CANADA) 6(9):27
Clemons, Gary (CDM Federal Programs Corp./USA) 6(6):81
Cocos, Ioana A. (École Polytechnique de Montréal/CANADA) 6(9):27
Cocucci, M. (Universita' degli Studi di Milano/ITALY) 6(5):157
Coelho, Rodrigo O. (CSD-GEOLOCK/BRAZIL) 6(1):27
Collet, Berto (TAUW bv/THE NETHERLANDS) 6(10):253
Compton, Joanne C. (REACT Environmental Engineers/USA) 6(3):25
Connell, Doug (Barr Engineering Company/USA) 6(5):105
Connor, Michael A. (University of Melbourne/AUSTRALIA) 6(10):329
Cook, Jim (Beazer East, Inc./USA) 6(2):239
Cooke, Larry (NOVA Chemicals Corporation/USA) 6(4):117
Coons, Darlene (Conestoga-Rovers & Associates/USA) 6(1):99; 6(10):131
Costley, Shauna C. (University of Natal/REP OF SOUTH AFRICA) 6(9):79
Cota, Jennine L. (ARCADIS Geraghty & Miller, Inc./USA) 6(7):149
Covell, James R. (EG&G Technical Services, Inc./USA) 6(10):49
Cowan, James D. (Ensafe Inc./USA) 6(9):315
Cox, Evan E. (GeoSyntec Consultants/CANADA) 6(8):27, 6(9):297
Cox, Jennifer (Clemson University/USA) 6(7):109
Craig, Shannon (Beazer East, Inc./USA) 6(2):239
Crawford, Donald L. (University of Idaho/USA) 6(3):91; 6(9):147

Crecelius, Eric (Battelle/USA) 6(5): 231, 237
Crotwell, Terry (Solutions Industrial & Environmental Services/USA) 6(7):249
Cui, Yanshan (Chinese Academy of Sciences/CHINA) 6(9):113
Cunningham, Al B. (Montana State University/USA) 6(7):117; 6(8):1
Cunningham, Jeffrey A. (Stanford University/USA) 6(7):95
Cutright, Teresa J. (The University of Akron/USA) 6(3):235

da Silva, Marcio Luis Busi (University of Iowa/USA) 6(1):195
Daly, Daniel J. (Energy & Environmental Research Center/USA) 6(5):129
Daniel, Fabien (AEA Technology Environment/UK) 6(10):337
Daniels, Gary (GeoTrans/USA) 6(8):19
Das, K.C. (University of Georgia/USA) 6(9):289
Davel, Jan L. (University of Cincinnati/USA) 6(6):133
Davis, Gregory A. (Microbial Insights Inc./USA) 6(2):97
Davis, Jeffrey L. (U.S. Army/USA) 6(3): 43, 51
Davis, John W. (The Dow Chemical Company/USA) 6(2):89
Davis-Hoover, Wendy J. (U.S. EPA/USA) 6(2):247
De'Ath, Anna M. (Cranfield University/UK) 6(6):329
Dean, Sean (Camp Dresser & McKee. Inc/USA) 6(7):133
DeBacker, Dennis (Battelle/USA) 6(10):145
DeHghi, Benny (Honeywell International Inc./USA) 6(2):39; 6(10):283
de Jong, Jentsje (TAUW BV/THE NETHERLANDS) 6(10):253
Del Vecchio, Michael (Envirogen, Inc./USA) 6(9):281
Delille, Daniel (CNRS/FRANCE) 6(2):57
DeLong, George (AIMTech/USA) 6(7):321, 333, 341
Demers, Gregg (ERM/USA) 6(7):45
De Mot, Rene (Catholic University of Leuven/BELGIUM) 6(4):35

Deobald, Lee A. (University of Idaho/USA) 6(9):147
Deschênes, Louise (École Polytechnique de Montréal/CANADA) 6(3):115; 6(9):129
Dey, William S. (Illinois State Geological Survey/USA) 6(9):179
Díaz-Cervantes, Dolores (CINVESTAV-IPN/MEXICO) 6(6):369
Dick, Vincent B. (Haley & Aldrich, Inc./USA) 6(7):27, 281
Diehl, Danielle (The University of New Mexico/USA) 6(9):155
Diehl, Susan V. (Mississippi State University/USA) 6(5):329
Diels, Ludo (VITO/BELGIUM) 6(9):87
DiGregorio, Salvatore (University della Calabria/ITALY) 6(4):131
Di Gregorio, Simona (Universita degli Studi di Verona/ITALY) 6(3):267
Dijkhuis, Edwin (Bioclear/THE NETHERLANDS) 6(5):289
Di Leo, Cristina (EniTecnologie/ITALY) 6(6):173
Dimitriou-Christidis, Petros (Texas A&M University) 6(5):17
Dixon, Robert (Montgomery Watson/ITALY) 6(10):41
Dobbs, Gregory M. (United Technologies Research Center/USA) 6(7):69
Doherty, Amy T. (GZA GeoEnvironmental, Inc./USA) 6(7):165
Dolan, Mark E. (Oregon State University/USA) 6(10):145, 155, 179
Dollhopf, Michael (Michigan State University/USA) 6(8):19
Dondi, Giovanni (Water & Soil Remediation S.r.l./ITALY) 6(6):179
Dong, Yiting (Chinese Academy of Sciences/CHINA) 6(9):113
Dooley, Maureen A. (Regenesis/USA) 6(7):197
Dottridge, Jane (Komex Europe Ltd./UK) 6(4):17
Dowd, John (University of Georgia/USA) 6(9):289
Doughty, Herb (U.S. Navy/USA) 6(10):1

Author Index

Doze, Jacco (RIZA/THE NETHERLANDS) 6(5):289
Dragich, Brian (California Polytechnic State University/USA) 6(2):1
Drake, John T. (Camp Dresser & McKee Inc./USA) 6(7):273
Dries, Victor (Flemish Public Waste Agency/BELGIUM) 6(7):87
Du, Yan-Hung (National Chung Hsing University/TAIWAN) 6(6):353
Dudal, Yves (École Polytechnique de Montréal/CANADA) 6(3):115
Duffey, J. Tom (Camp Dresser & McKee Inc./USA) 6(5):33
Duffy, Baxter E. (Inland Pollution Services, Inc./USA) 6(7):313
Duijn, Rik (Oostwaardhoeve Co./THE NETHERLANDS) 6(6):59
Durant, Neal D. (GeoTrans, Inc./USA) 6(2):19, 163
Durell, Gregory (Battelle Ocean Sciences/USA) 6(5):231
Dworatzek, S. (University of Toronto/CANADA) 6(8):27
Dwyer, Daryl F. (University of Minnesota/USA) 6(3):219
Dzantor, E. K. (University of Maryland/USA) 6(5):77

Ebner, R. (GMF/GERMANY) 6(2):137
Ederer, Martina (University of Idaho/USA) 6(9):147
Edgar, Michael (Camp Dresser & McKee Inc./USA) 6(7):133
Edwards, Elizabeth A. (University of Toronto/CANADA) 6(8):27
Edwards, Grant C. (University of Guelph/CANADA) 6(6):185
Eggen, Trine (Jordforsk Centre for Soil and Environmental Research/NORWAY) 6(6):157
Eggert, Tim (CDM Federal Programs Corp./USA) 6(6):81
Elberson, Margaret A. (DuPont Co./USA) 6(8):43
Elliott, Mark (Virginia Polytechnic Institute & State University/USA) 6(5):1
Ellis, David E. (Dupont Company/USA) 6(8):43

Ellwood, Derek C. (University of Southampton/UK) 6(9):61
Else, Terri (University of Nevada Las Vegas/USA) 6(9):257
Elväng, Annelie M. (Stockholm University/SWEDEN) 6(3):133
England, Kevin P. (USA) 6(5):105
Ertas, Tuba Turan (San Diego State University/USA) 6(10):301
Escalon, Lynn (U.S. Army Corps of Engineers/USA) 6(3):51
Esparza-Garcia, Fernando (CINVESTAV-IPN/MEXICO) 6(6):219
Evans, Christine S. (University of Westminster/UK) 6(3):75
Evans, Patrick J. (Camp Dresser & McKee, Inc./USA) 6(2):113, 199; 6(8):209

Fabiani, Fabio (EniTecnologie S.p.A./ITALY) 6(6):173
Fadullon, Frances Steinacker (CH2M Hill/USA) 6(3):107
Fang, Min (University of Massachusetts/USA) 6(6):73
Faris, Bart (New Mexico Environmental Department/USA) 6(9):223
Farone, William A. (Applied Power Concepts, Inc./USA) 6(7):103
Fathepure, Babu Z. (Oklahoma State University/USA) 6(8):19
Faust, Charles (GeoTrans, Inc./USA) 6(2):163
Fayolle, Françoise (Institut Français du Pétrole/FRANCE) 6(1):153
Feldhake, David (University of Cincinnati/USA) 6(2):247
Felt, Deborah (Applied Research Associates, Inc./USA) 6(7):125
Feng, Terry H. (Parsons Engineering Science, Inc./USA) 6(2):39; 6(10):283
Fenwick, Caroline (Aberdeen University/UK) 6(2):223
Fernandez, Jose M. (University of Iowa/USA) 6(1):195
Fernández-Sanchez, J. Manuel (CINVESTAV-IPN/MEXICO) 6(6):369

Ferrer, E. (Universidad Complutense de Madrid/SPAIN) 6(4):29
Ferrera-Cerrato, Ronald (Colegio de Postgraduados/MEXICO) 6(6):219
Fiacco, R. Joseph (Environmental Resources Management) 6(7):45
Fields, Jim (University of Georgia/USA) 6(9):289
Fields, Keith A. (Battelle/USA) 6(10):1
Fikac, Paul J. (Jacobs Engineering Group, Inc./USA) 6(6):35
Fischer, Nick M. (Aquifer Technology/USA) 6(8):157, 6(10):15
Fisher, Angela (The Pennsylvania State University/USA) 6(8):201
Fisher, Jonathan (Environment Agency/UK) 6(4):17
Fitch, Mark W. (University of Missouri-Rolla/USA) 6(5):199
Fleckenstein, Janice V. (USA) 6(6):89
Fleischmann, Paul (ZEBRA Environmental Corp./USA) 6(10):139
Fletcher, John S. (University of Oklahoma/USA) 6(5):61
Foget, Michael K. (SHN Consulting Engineers & Geologists, Inc./USA) 6(3):83
Foley, K.L. (U.S. Army Engineer Research & Development Center/USA) 6(5):9
Follner, Christina G. (University of Leipzig/GERMANY) 6(4):81
Fontenot, Martin M. (Syngenta Crop Protection, Inc./USA) 6(6):35
Foote, Eric A. (Battelle/USA) 6(1):137; 6(7):13
Ford, James (Investigative Science Inc./CANADA) 6(2):27
Forman, Sarah R. (URS Corporation/USA) 6(7):321, 333, 341
Fortman, Tim J. (Battelle Marine Sciences Laboratory/USA) 6(3):157
Francendese, Leo (U.S. EPA/USA) 6(3):259
Francis, M. McD. (NOVA Research & Technology Center/CANADA) 6(4):117; 6(5):53,
François, Alan (Institut Français du Pétrole/FRANCE) 6(1):153

Frankenberger, William T. (University of California/USA) 6(9):249
Freedman, David L. (Clemson University/USA) 6(7):109
French, Christopher E. (University of Cambridge/UK) 6(5):69
Friese, Kurt (UFZ Center for Environmental Research/GERMANY) 6(9):43
Frisbie, Andrew J. (Purdue University/USA) 6(3):125
Frisch, Sam (Envirogen Inc./USA) 6(9):281
Frömmichen, René (UFZ Centre for Environmental Research/GERMANY) 6(9):43
Fuierer, Alana M. (New Mexico Institute of Mining & Technology/USA) 6(8):131
Fujii, Kensuke (Obayashi Corporation/JAPAN) 6(10):239
Fujii, Shigeo (Kyoto University/JAPAN) 6(4):149
Furuki, Masakazu (Hyogo Prefectural Institute of Environmental Science/JAPAN) 6(5):321

Gallagher, John R. (University of North Dakota/USA) 6(5):129; 6(6):141
Gambale, Franco (Istituto di Cibernetica e Biofisica/ITALY) 6(5):157
Gambrell, Robert P. (Louisiana State University/USA) 6(5):305
Gandhi, Sumeet (University of Iowa/USA) 6(8):147
Garbi, C. (Universidad Complutense de Madrid/SPAIN) 6(4):29; 6(6):377
García-Arrazola, Roeb (CINVESTAV-IPN/MEXICO) 6(6):369
García-Barajas, Rubén Joel (ESIQIE-IPN/MEXICO) 6(6):369
Garrett, Kevin (Harding ESE/USA) 6(7):205
Garry, Erica (Spelman College/USA) 6(9):53
Gavaskar, Arun R. (Battelle/USA) 6(7):13
Gavinelli, Marco (Ambiente S.p.A./ITALY) 6(6):227
Gebhard, Michael (GeoTrans/USA) 6(8):19

Gec, Bob (Degussa Canada Ltd./CANADA) 6(10):73
Gehre, Matthias (UFZ - Centre for Environmental Research/GERMANY) 6(4):99
Gemoets, Johan (VITO/BELGIUM) 6(4):35; 6(9):87
Gent, David B. (U.S. Army Corps of Engineers/USA) 6(9):241
Gentry, E. E. (Science Applications International Corporation/USA) 6(8):27
Georgiev, Plamen S. (University of Mining & Geology/BULGARIA) 6(9):97
Gerday, Charles (Université de Liège/BELGIUM) 6(2):57
Gerlach, Robin (Montana State University/USA) 6(8):1
Gerritse, Jan (TNO Environmental Sciences/THE NETHERLANDS) 6(2):231; 6(7):61
Gerth, André (BioPlanta GmbH/GERMANY) 6(3):67; 6(5):173
Ghosh, Upal (Stanford University/USA) 6(3):189; 6(6):89
Ghoshal, Subhasis (McGill University/CANADA) 6(9):139
Gibbs, James T. (Battelle/USA) 6(1):137
Gibello, A. (Universidad Complutense/SPAIN) 6(4):29
Giblin, Tara (University of California/USA) 6(9):249
Gilbertson, Amanda W. (University of Missouri-Rolla/USA) 6(5):199
Gillespie, Rick D. (Regenesis/USA) 6(1):107
Gillespie, Terry J. (University of Guelph/CANADA) 6(6):185
Glover, L. Anne (Aberdeen University /UK) 6(2):223
Goedbloed, Peter (Oostwaardhoeve Co./THE NETHERLANDS) 6(6):59
Golovleva, Ludmila A. (Russian Academy of Sciences/RUSSIA) 6(3):75
Goltz, Mark N. (Air Force Institute of Technology/USA) 6(2):173

Gong, Weiliang (The University of New Mexico/USA) 6(9):155
Gossett, James M. (Cornell University/USA) 6(4):125
Govind, Rakesh (University of Cincinnati/USA) 6(5):269; 6(8):35; 6(9):1, 9, 17
Gozan, Misri (Water Technology Center/GERMANY) 6(8):105
Grainger, David (IT Corporation/USA) 6(1):51; 6(2):73
Grandi, Beatrice (Water & Soil Remediation S.r.l./ITALY) 6(6):179
Granley, Brad A. (Leggette, Brashears, & Graham/USA) 6(10):259
Grant, Russell J. (University of York/UK) 6(6):27
Graves, Duane (IT Corporation/USA) 6(2):253; 6(4):109; 6(9):215
Green, Chad E. (University of California/USA) 6(10):311
Green, Donald J. (USAG Aberdeen Proving Ground/USA) 6(7):321, 333, 341
Green, Robert (Alcoa/USA) 6(6):89
Green, Roger B. (Waste Management, Inc./USA) 6(2):247; 6(6):127
Gregory, Kelvin B. (University of Iowa/USA) 6(3):1
Griswold, Jim (Construction Analysis & Management, Inc./USA) 6(1):115
Groen, Jacobus (Vrije Universiteit/THE NETHERLANDS) 6(4):91
Groenendijk, Gijsbert Jan (Hoek Loos bv/THE NETHERLANDS) 6(7):297
Grotenhuis, Tim (Wageningen Agricultural University/THE NETHERLANDS) 6(5):289
Groudev, Stoyan N. (University of Mining & Geology/BULGARIA) 6(9):97
Guarini, William J. (Envirogen, Inc./USA) 6(9):281
Guieysse, Benoît (Lund University/SWEDEN) 6(3):181
Guiot, Serge R. (Biotechnology Research Institute/CANADA) 6(3):165
Gunsch, Claudia (Clemson University/USA) 6(7):109
Gurol, Mirat (San Diego State University/USA) 6(10):301

Ha, Jeonghyub (University of Maryland/USA) 6(10):57
Haak, Daniel (RMT, Inc./USA) 6(10):201
Haas, Patrick E. (Mitretek Systems/USA) 6(7):19, 241, 249; 6(8):73
Haasnoot, C. (Logisticon Water Treatment/THE NETHERLANDS) 6(8):11
Habe, Hiroshi (The University of Tokyo/JAPAN) 6(4):51; 6(6):111
Haeseler, Frank (Institut Français du Pétrole/FRANCE) 6(3):227
Haff, James (Meritor Automotive, Inc./USA) 6(7):173
Haines, John R. (U.S. EPA/USA) 6(9):17
Håkansson, Torbjörn (Lund University/SWEDEN) 6(9):123
Halfpenny-Mitchell, Laurie (University of Guelph/CANADA) 6(6):185
Hall, Billy (Newfields, Inc./USA) 6(5):189
Hampton, Mark M. (Groundwater Services/USA) 6(8):73
Hannick, Nerissa K. (University of Cambridge/UK) 6(5):69
Hannigan, Mary (Mississippi State University) 6(5):329; 6(6):279
Hannon, LaToya (Spelman College/USA) 6(9):53
Hansen, Hans C. L. (Hedeselskabet /DENMARK) 6(2):11
Hansen, Lance D. (U.S. Army Corps of Engineers/USA) 6(3):9, 43, 51; 6(4):59; 6(6):43; 6(7):125; 6(10):115
Haraguchi, Makoo (Sumitomo Marine Research Institute/JAPAN) 6(10):345
Hardisty, Paul E. (Komex Europe, Ltd./ENGLAND) 6(4):17
Harmon, Stephen M. (U.S. EPA/USA) 6(9):17
Harms, Hauke (Swiss Federal Institute of Technology/SWITZERLAND) 6(3):251
Harmsen, Joop (Alterra, Wageningen University and Research Center/THE NETHERLANDS) 6(5):137, 279; 6(6):1, 59

Harper, Greg (TetraTech EM Inc./USA) 6(3):259
Harrington-Baker, Mary Ann (MSE, Inc./USA) 6(9):35
Harris, Benjamin Cord (Texas A&M University/USA) 6(5):17, 25
Harris, James C. (U.S. EPA/USA) 6(6):287, 295
Harris, Todd (Mason and Hanger Corporation/USA) 6(3):35
Harrison, Patton B. (American Airlines/USA) 6(1):121
Harrison, Susan T.L. (University of Cape Town/REP OF SOUTH AFRICA) 6(6):339
Hart, Barry (Monash University/AUSTRALIA) 6(4):1
Hartzell, Kristen E. (Battelle/USA) 6(1):137; 6(10):193
Harwood, Christine L. (Michael Baker Corporation/USA) 6(2):155
Hassett, David J. (Energy & Environmental Research Center/USA) 6(5):129
Hater, Gary R. (Waste Management Inc./USA) 6(2):247
Hausmann, Tom S. (Battelle Marine Sciences Laboratory/USA) 6(3):157
Hawari, Jalal (National Research Council of Canada/CANADA) 6(9):139
Hayes, Adam J. (Triple Point Engineers, Inc./USA) 6(1):183
Hayes, Dawn M. (U.S. Navy/USA) 6(3):107
Hayes, Kim F. (University of Michigan/USA) 6(8):193
Haynes, R.J. (University of Natal/REP OF SOUTH AFRICA) 6(6):101
Heaston, Mark S. (Earth Tech/USA) 6(3):17, 25
Hecox, Gary R. (University of Kansas/USA) 6(4):109
Heebink, Loreal V. (Energy & Environmental Research Center/USA) 6(5):129
Heine, Robert (EFX Systems, Inc./USA) 6(8):19
Heintz, Caryl (Texas Tech University/USA) 6(3):9

Hendrickson, Edwin R. (DuPont Co./USA) 6(8):27, 43
Hendriks, Willem (Witteveen+Bos Consulting Engineers/THE NETHERLANDS) 6(5):289
Henkler, Rolf D. (ICI Paints/UK) 6(2):223
Henny, Cynthia (University of Maine/USA) 6(8):139
Henry, Bruce M. (Parsons Engineering Science, Inc/USA) 6(7):241
Henssen, Maurice J.C. (Bioclear Environmental Biotechnology/THE NETHERLANDS) 6(8):11
Herson, Diane S. (University of Delaware/USA) 6(6):9
Hesnawi, Rafik M. (University of Manitoba/CANADA) 6(6):165
Hetland, Melanie D. (Energy & Environmental Research Center/USA) 6(5):129
Hickey, Robert F. (EFX Systems, Inc./USA) 6(8):19
Hicks, Patrick H. (ARCADIS/USA) 6(1):107
Hiebert, Randy (MSE Technology Applications, Inc./USA) 6(8):79
Higashi, Teruo (University of Tsukuba/JAPAN) 6(9):187
Higgins, Mathew J. (Bucknell University/USA) 6(2):105
Higinbotham, James H. (ExxonMobil Environmental Remediation/USA) 6(8):87
Hines, April (Spelman College/USA) 6(9):53
Hinshalwood, Gordon (Delta Environmental Consultants, Inc./USA) 6(1):43
Hirano, Hiroyuki (The University of Tokyo/JAPAN) 6(6):111
Hirashima, Shouji (Yakult Pharmaceutical Industry/JAPAN) 6(10):345
Hirsch, Steve (Environmental Protection Agency/USA) 6(5):207
Hiwatari, Takehiko (National Institute for Environmental Studies/JAPAN) 6(5):321
Hoag, Rob (Conestoga-Rovers & Associates/USA) 6(1):99

Hoelen, Thomas P. (Stanford University/USA) 6(7):95
Hoeppel, Ronald E. (U.S. Navy/USA) 6(10):245
Hoffmann, Johannes (Hochtief Umwelt GmbH/GERMANY) 6(6):227
Hoffmann, Robert E. (Chevron Canada Resources/CANADA) 6(6):193
Höfte, Monica (Ghent University/BELGIUM) 6(5):223
Holder, Edith L. (University of Cincinnati/USA) 6(2):247
Holm, Thomas R. (Illinois State Water Survey/USA) 6(9):179
Holman, Hoi-Ying (Lawrence Berkeley National Laboratory/USA) 6(4):67
Holoman, Tracey R. Pulliam (University of Maryland/USA) 6(3):205
Hopper, Troy (URS Corporation/USA) 6(2):239
Hornett, Ryan (NOVA Chemicals Corporation/USA) 6(4):117
Hosangadi, Vitthal S. (Foster Wheeler Environmental Corp./USA) 6(9):249
Hough, Benjamin (Tetra Tech EM, Inc./USA) 6(10):293
Hozumi, Toyoharu (Oppenheimer Biotechnology/JAPAN) 6(10):345
Huang, Chin-I (National Chung Hsing University/TAIWAN) 6(10):217
Huang, Chin-Pao (University of Delaware/USA) 6(6):9, 149
Huang, Hui-Bin (DuPont Co./USA) 6(8):43
Huang, Junqi (Air Force Institute of Technology/USA) 6(2):173
Huang, Wei (University of Sheffield/UK) 6(2):207
Hubach, Cor (DHV Noord Nederland/THE NETHERLANDS) 6(8):11
Huesemann, Michael H. (Battelle/USA) 6(3):157
Hughes, Joseph B. (Rice University/USA) 6(5):85; 6(7):19
Hulsen, Kris (University of Ghent/BELGIUM) 6(5):223
Hunt, Jonathan (Clemson University/USA) 6(7):109

Hunter, William J. (U.S. Dept of Agriculture/USA) 6(9):209, 309
Hwang, Sangchul (University of Akron/USA) 6(3):235
Hyman, Michael R. (North Carolina State University/USA) 6(1): 83, 145

Ibeanusi, Victor M. (Spelman College/USA) 6(9):53
Ickes, Jennifer (Battelle/USA) 6(5):231, 237
Ide, Kazuki (Obayashi Corporation Ltd./JAPAN) 6(6):111; 6(10):239
Igarashi, Tsuyoshi (Nippon Institute of Technology/JAPAN) 6(5):321
Infante, Carmen (PDVSA Intevep/VENEZUELA) 6(6):257
Ingram, Sherry (IT Corporation/USA) 6(4):109
Ishikawa, Yoji (Obayashi Corporation/JAPAN) 6(6):249; 6(10):239

Jackson, W. Andrew (Texas Tech University/USA) 6(5):207, 313; 6(9):273
Jacobs, Alan K. (EnSafe, Inc./USA) 6(9):315
Jacques, Margaret E. (Rowan University/USA) 6(5):215
Jahan, Kauser (Rowan University/USA) 6(5):215
James, Garth (MSE Inc./USA) 6(8):79
Jansson, Janet K. (Södertörn University College/SWEDEN) 6(3):133
Japenga, Jan (Alterra/THE NETHERLANDS) 6(5):137
Jauregui, Juan (Universidad Nacional Autonoma de Mexico/MEXICO) 6(6):17
Jensen, James N. (State University of New York at Buffalo/USA) 6(6):89
eon, Mi-Ae (Texas Tech University/USA) 6(9):273
Jerger, Douglas E. (IT Corporation/USA) 6(3):35
Jernberg, Cecilia (Södertörn University College/SWEDEN) 6(3):133
Jindal, Ranjna (Suranaree University of Technology/THAILAND) 6(4):149

Johnson, Dimitra (Southern University at New Orleans/USA) 6(5):151
Johnson, Glenn (University of Utah/USA) 6(5):231
Johnson, Paul C. (Arizona State University/USA) 6(1):11; 6(8):61
Johnson, Richard L. (Oregon Graduate Institute/USA) 6(10):293
Jones, Antony (Komex H_2O Science, Inc./USA) 6(2):223; 6(3):173; 6(10):123
Jones, Clay (University of New Mexico/USA) 6(9):223
Jones, Triana N. (University of Maryland/USA) 6(3):205
Jonker, Hendrikus (Vrije Universiteit/THE NETHERLANDS) 6(4):91
Ju, Lu-Kwang (The University of Akron/USA) 6(6):319

Kaludjerski, Milica (San Diego State University/USA) 6(10):301
Kamashwaran, S. Ramanathen (University of Idaho/USA) 6(3):91
Kambhampati, Murty S. (Southern University at New Orleans/USA) 6(5):145, 151
Kamimura, Daisuke (Gunma University/JAPAN) 6(8):113
Kang, James J. (URS Corporation/USA) 6(1):121; 6(10):223
Kappelmeyer, Uwe (UFZ Centre for Environmental Research/GERMANY) 6(5):337
Karamanev, Dimitre G. (University of Western Ontario/CANADA) 6(10):171
Karlson, Ulrich (National Environmental Research Institute) 6(3):141
Kastner, James R. (University of Georgia/USA) 6(9):289
Kästner, Matthias (UFZ Centre for Environmental Research/GERMANY) 6(4):99; 6(5):337
Katz, Lynn E. (University of Texas/USA) 6(8):139
Kavanaugh, Rathi G. (University of Cincinnati/USA) 6(2):247

Kawahara, Fred (U.S. EPA/USA) 6(9):9
Kawakami, Tsuyoshi (University of Tsukuba/JAPAN) 6(9):187
Keefer, Donald A. (Illinois State Geological Survey/USA) 6(9):179
Keith, Nathaniel (Texas A&M University/USA) 6(5):25
Kelly, Laureen S. (Montana Department of Environmental Quality/USA) 6(6):287
Kempisty, David M. (U.S. Air Force/USA) 6(10):145, 155
Kerfoot, William B. (K-V Associates, Inc./USA) 6(10):33
Keuning, S. (Bioclear Environmental Technology/THE NETHERLANDS) 6(8):11
Khan, Tariq A. (Groundwater Services, Inc./USA) 6(7):19
Khodadoust, Amid P. (University of Cincinnati/USA) 6(5):243, 253, 261
Kieft, Thomas L. (New Mexico Institute of Mining and Technology/USA) 6(8):131
Kiessig, Gunter (WISMUT GmbH/GERMANY) 6(5):173; 6(9):155
Kilbride, Rebecca (CEFAS Laboratory/UK) 6(10):337
Kim, Jae Young (Seoul National University/REPUBLIC OF KOREA) 6(9):195
Kim, Jay (University of Cincinnati/USA) 6(6):133
Kim, Kijung (The Pennsylvania State University/USA) 6(9):303
Kim, Tae Young (Ewha University/REPUBLIC OF KOREA) 6(6):51
Kinsall, Barry L. (Oak Ridge National Laboratory/USA) 6(4):73
Kirschenmann, Kyle (IT Corp/USA) 6(4):109
Klaas, Norbert (University of Stuttgart/GERMANY) 6(2):137
Klecka, Gary M. (The Dow Chemical Company/USA) 6(2):89
Klein, Katrina (GeoTrans, Inc./USA) 6(2):163

Klens, Julia L. (IT Corporation/USA) 6(2):253; 6(9):215
Knotek-Smith, Heather M. (University of Idaho/USA) 6(9):147
Koch, Stacey A. (RMT, Inc./USA) 6(7):181
Koenen, Brent A. (U.S. Army Engineer Research & Development Center/USA) 6(5):9
Koenigsberg, Stephen S. (Regenesis Bioremediation Products/USA) 6(7):197, 257; 6(8):209; 6(10):9, 87
Kohata, Kunio (National Institute for Environmental Studies/JAPAN) 6(5):321
Kohler, Keisha (ThermoRetec Corporation/USA) 6(7):1
Kolhatkar, Ravindra V. (BP Corporation/USA) 6(1):35, 43
Komlos, John (Montana State University/USA) 6(7):117
Komnitsas, Kostas (National Technical University of Athens/GREECE) 6(9):97
Kono, Masakazu (Oppenheimer Biotechnology/JAPAN) 6(10):345
Koons, Brad W. (Leggette, Brashears & Graham, Inc./USA) 6(1):175
Koschal, Gerard (PNG Environmental/USA) 6(1):203
Koschorreck, Matthias (UFZ Centre for Environmental Research/GERMANY) 6(9):43
Koshikawa, Hiroshi (National Institute for Environmental Studies/JAPAN) 6(5):321
Kramers, Jan D. (University of Bern/SWITZERLAND) 6(4):91
Krooneman, Jannneke (Bioclear Environmental Biotechnology/THE NETHERLANDS) 6(7):141
Kruk, Taras B. (URS Corporation/USA) 6(10):223
Kuhwald, Jerry (NOVA Chemicals Corporation/CANADA) 6(5):53
Kuschk, Peter (UFZ Centre for Environmental Research Leipzig/GERMANY) 6(5):337

Laboudigue, Agnes (Centre National de Recherche sur les Sites et Sols Pollués/FRANCE) 6(2):129
LaFlamme, Brian (Engineering Management Support, Inc./USA) 6(10):231
Lafontaine, Chantal (École Polytechnique de Montréal/CANADA) 6(10):171
Laha, Shonali (Florida International University/USA) 6(10):187
Laing, M.D. (University of Natal/REP OF SOUTH AFRICA) 6(9):79
Lamar, Richard (EarthFax Development Corp/USA) 6(6):263
Lamarche, Philippe (Royal Military College of Canada/CANADA) 6(8):95
Lamb, Steven R. (GZA GeoEnvironmental, Inc./USA) 6(7):165
Landis, Richard C. (E.I. du Pont de Nemours & Company/USA) 6(8):185
Lang, Beth (United Technologies Corp./USA) 6(10):41
Langenhoff, Alette (TNO Institute of Environmental Science/THE NETHERLANDS) 6(7):141
LaPat-Polasko, Laurie T. (Parsons Engineering Science, Inc./USA) 6(2):65, 189
Lapus, Kevin (Regenesis/USA) 6(7):257; 6(10):9
LaRiviere, Daniel (Texas A&M University/USA) 6(5):17, 25
Larsen, Lars C. (Hedeselskabet/DENMARK) 6(2):11
Larson, John R. (TranSystems Corporation/USA) 6(7):229
Larson, Richard A. (University of Illinois at Urbana-Champaign/USA) 6(5):181
Lauzon, Francois (Dept of National Defence/CANADA) 6(8):95
Leavitt, Maureen E. (Newfields Inc./USA) 6(1):51; 6(5):189
Lebron, Carmen A. (U.S. Navy/USA) 6(7):95
Lee, B. J. (Science Applications International Corporation) 6(8):27

Lee, Brady D. (Idaho National Engineering & Environmental Laboratory/USA) 6(7):77
Lee, Chi Mei (National Chung Hsing University/TAIWAN) 6(6):353
Lee, Eun-Ju (Louisiana State University/USA) 6(5):313
Lee, Kenneth (Fisheries & Oceans Canada/CANADA) 6(10):337
Lee, Michael D. (Terra Systems, Inc./USA) 6(7):213, 249
Lee, Ming-Kuo (Auburn University/USA) 6(9):105
Lee, Patrick (Queen's University/CANADA) 6(2):121
Lee, Seung-Bong (University of Washington/USA) 6(10):211
Lee, Si-Jin (Kyonggi University/REPUBLIC OF KOREA) 6(1):161
Lee, Sung-Jae (ChoongAng University/REPUBLIC OF KOREA) 6(6):51
Leeson, Andrea (Battelle/USA) 6(10):1, 145, 155, 193
Lehman, Stewart E. (California Polytechnic State University/USA) 6(2):1
Lei, Li (University of Cincinnati/USA) 6(5):243, 261
Leigh, Daniel P. (IT Corporation/USA) 6(3):35
Leigh, Mary Beth (University of Oklahoma/USA) 6(5):61
Lendvay, John (University of San Francisco/USA) 6(8):19
Lenzo, Frank C. (ARCADIS Geraghty & Miller/USA) 6(7):53
Leon, Nidya (PDVSA Intevep/VENEZUELA) 6(6):257
Leong, Sylvia (Crescent Heights High School/CANADA) 6(5):53
Leontievsky, Alexey A. (Russian Academy of Sciences/RUSSIA) 6(3):75
Lerner, David N. (University of Sheffield/UK) 6(1):59; 6(2):207
Lesage, Suzanne (National Water Research Institute/CANADA) 6(7):321, 333, 341

Leslie, Jolyn C. (Camp Dresser & McKee, Inc./USA) 6(2):113
Lewis, Ronald F. (U.S. EPA/USA) 6(5):253, 261
Li, Dong X. (USA) 6(7):205
Li, Guanghe (Tsinghua University/CHINA) 6(7):61
Li, Tong (Tetra Tech EM Inc./USA) 6(10):293
Librando, Vito (Universita di Catania/ITALY) 6(3):149
Lieberman, M. Tony (Solutions Industrial & Environmental Services/USA) 6(7):249
Lin, Cindy (Conestoga-Rovers & Associates/USA) 6(1):99; 6(10):131
Lipson, David S. (Blasland, Bouck & Lee, Inc./USA) 6(10):319
Liu, Jian (University of Nevada Las Vegas/USA) 6(9):265
Liu, Xiumei (Shandong Agricultural University/ CHINA) 6(9):113
Livingstone, Stephen (Franz Environmental Inc./CANADA) 6(6):211
Lizzari, Daniela (Universita degli Studi di Verona/ITALY) 6(3):267
Llewellyn, Tim (URS/USA) 6(7):321, 333, 341
Lobo, C. (El Encin IMIA/SPAIN) 6(4):29
Loeffler, Frank E. (Georgia Institute of Technology/USA) 6(8):19
Logan, Bruce E. (The Pennsylvania State University/USA) 6(9):303
Long, Gilbert M. (Camp Dresser & McKee Inc./USA) 6(6):287
Longoni, Giovanni (Montgomery Watson/ITALY) 6(10):41
Lorbeer, Helmut (Technical University of Dresden/GERMANY) 6(8):105
Lors, Christine (Centre National de Recherche sur les Sites et Sols Pollués /FRANCE) 6(2):129
Lorton, Diane M. (King's College London/UK) 6(2):223; 6(3):173
Losi, Mark E. (Foster Wheeler Environ. Corp./USA) 6(9):249
Loucks, Mark (U.S. Air Force/USA) 6(2):261

Lu, Chih-Jen (National Chung Hsing University/TAIWAN) 6(6):353; 6(10):217
Lu, Xiaoxia (Tsinghua University/CHINA) 6(7):61
Lubenow, Brian (University of Delaware/USA) 6(6):149
Lucas, Mary (Parsons Engineering Science, Inc./USA) 6(10):283
Lundgren, Tommy S. (Sydkraft SAKAB AB/SWEDEN) 6(6):127
Lundstedt, Staffan (Umeå University/SWEDEN) 6(3):181
Luo, Xiaohong (NRC Research Associate/USA) 6(8):167
Luthy, Richard G. (Stanford University/USA) 6(3):189
Lutze, Werner (University of New Mexico/USA) 6(9):155
Luu, Y.-S. (Queen's University/CANADA) 6(2):121
Lynch, Regina M. (Battelle/USA) 6(10):155

Macek, Thomáš (Institute of Chemical Technology/Czech Republic) 6(5):61
MacEwen, Scott J. (CH2M Hill/USA) 6(3):107
Machado, Kátia M. G. (Fund. Centro Tecnológico de Minas Gerais/BRAZIL) 6(3):99
Maciel, Helena Alves (Aberdeen University/UK) 6(1):1
Mack, E. Erin (E.I. du Pont de Nemours & Co./USA) 6(2):81; 6(8):43
Macková, Martina (Institute of Chemical Technology/Czech Republic) 6(5):61
Macnaughton, Sarah J. (AEA Technology/UK) 6(5):305; 6(10):337
Macomber, Jeff R. (University of Cincinnati/USA) 6(6):133
Macrae, Jean (University of Maine/USA) 6(8):139
Madden, Patrick C. (Engineering Consultant/USA) 6(8):87
Madsen, Clint (Terracon/USA) 6(8):157; 6(10):15
Magar, Victor S. (Battelle/USA) 6(1):137; 6(5):231, 237; 6(10):145, 155

Mage, Roland (Battelle Europe/SWITZERLAND) 6(6):241; 6(10):109
Magistrelli, P. (Istituto di Cibernetica e Biofisica/ITALY) 6(5):157
Maierle, Michael S. (ARCADIS Geraghty & Miller, Inc./USA) 6(7):149
Major, C. Lee (Jr.) (University of Michigan/USA) 6(8):19
Major, David W. (GeoSyntec Consultants/CANADA) 6(8):27
Maki, Hideaki (National Institute for Environmental Studies/JAPAN) 6(5):321
Makkar, Randhir S. (University of Illinois-Chicago/USA) 6(5):297
Malcolm, Dave (BAE Systems Properties Ltd./UK) 6(6):119
Manabe, Takehiko (Hyogo Prefectural Fisheries Research Institute/JAPAN) 6(10):345
Maner, P.M. (Equilon Enterprises, LLC/USA) 6(1):11
Maner, Paul (Shell Development Company/USA) 6(8):61
Manrique-Ramírez, Emilio Javier (SYMCA, S.A. de C.V./MEXICO) 6(6):369
Marchal, Rémy (Institut Français du Pétrole/FRANCE) 6(1):153
Maresco, Vincent (Groundwater & Environmental Srvcs/USA) 6(10):101
Marnette, Emile C. (TAUW BV/THE NETHERLANDS) 6(7):297
Marshall, Timothy R. (URS Corporation/USA) 6(2):49
Martella, L. (Istituto di Cibernetica e Biofisica/ITALY) 6(5):157
Martin, C. (Universidad Politecnica/SPAIN) 6(4):29
Martin, Jennifer P. (Idaho National Engineering & Environmental Laboratory/USA) 6(7):265
Martin, John F. (U.S. EPA/USA) 6(2):247
Martin, Margarita (Universidad Complutense de Madrid/SPAIN) 6(4):29; 6(6):377

Martinez-Inigo, M.J. (El Encin IMIA/SPAIN) 6(4):29
Martino, Lou (Argonne National Laboratory/USA) 6(5):207
Mascarenas, Tom (Environmental Chemistry/USA) 6(8):157
Mason, Jeremy (King's College London/UK) 6(2):223; 6(3):173; 6(10):123
Massella, Oscar (Universita degli Studi di Verona/ITALY) 6(3):267
Matheus, Dacio R. (Instituto de Botânica/BRAZIL) 6(3):99
Matos, Tania (University of Puerto Rico at Rio Piedras/USA) 6(9):179
Matsubara, Takashi (Obayashi Corporation/JAPAN) 6(6):249
Mattiasson, Bo (Lund University/SWEDEN) 6(3):181; 6(6):65; 6(9):123
McCall, Sarah (Battelle/USA) 6(10):155, 245
McCarthy, Kevin (Battelle Duxbury Operations/USA) 6(5):9
McCartney, Daryl M. (University of Manitoba/CANADA) 6(6):165
McCormick, Michael L. (The University of Michigan/USA) 6(8):193
McDonald, Thomas J. (Texas A&M University) 6(5):17
McElligott, Mike (U.S. Air Force/USA) 6(1):51
McGill, William B. (University of Northern British Columbia/CANADA) 6(4):7
McIntosh, Heather (U.S. Army/USA) 6(7):321, 333
McLinn, Eugene L. (RMT, Inc./USA) 6(5):121
McLoughlin, Patrick W. (Microseeps Inc./USA) 6(1):35
McMaster, Michaye (GeoSyntec Consultants/CANADA) 6(8):27, 43; 6(9):297
McMillen, Sara J. (Chevron Research & Technology Company/USA) 6(6):193
Meckenstock, Rainer U. (University of Tübingen/GERMANY) 6(4):99
Mehnert, Edward (Illinois State Geological Survey/USA) 6(9):179

Author Index

Meigio, Jodette L. (Idaho National Engineering & Environmental Laboratory/USA) 6(7):77
Meijer, Harro A.J. (University of Groningen/THE NETHERLANDS) 6(4):91
Meijerink, E. (Province of Drenthe/THE NETHERLANDS) 6(8):11
Merino-Castro, Glicina (Inst Technol y de Estudios Superiores/MEXICO) 6(6):377
Messier, J.P. (U.S. Coast Guard/USA) 6(1):107
Meyer, Michael (Environmental Resources Management/BELGIUM) 6(7):87
Meylan, S. (Queen's University/CANADA) 6(2):121
Miles, Victor (Duracell Inc./USA) 6(7):87
Millar, Kelly (National Water Research Institute/CANADA) 6(7):321, 333, 341
Miller, Michael E. (Camp Dresser & McKee, Inc./USA) 6(7):273
Miller, Thomas Ferrell (Lockheed Martin/USA) 6(3):259
Mills, Heath J. (Georgia Institute of Technology/USA) 6(9):165
Millward, Rod N. (Louisiana State University/USA) 6(5):305
Mishra, Pramod Chandra (Sambalpur University/INDIA) 6(9):173
Mitchell, David (AEA Technology Environment/UK) 6(10):337
Mitraka, Maria (Serres/GREECE) 6(6):89
Mocciaro, PierFilippo (Ambiente S.p.A./ITALY) 6(6):227
Moeri, Ernesto N. (CSD-GEOKLOCK/BRAZIL) 6(1):27
Moir, Michael (Chevron Research & Technology Co./USA) 6(1):83
Molinari, Mauro (AgipPetroli S.p.A/ITALY) 6(6):173
Mollea, C. (Politecnico di Torino/ITALY) 6(3):211
Mollhagen, Tony (Texas Tech University/USA) 6(3):9
Monot, Frédéric (Institut Français du Pétrole/FRANCE) 6(1):153

Moon, Hee Sun (Seoul National University/REPUBLIC OF KOREA) 6(9):195
Moosa, Shehnaaz (University of Cape Town/REP OF SOUTH AFRICA) 6(6):339
Morasch, Barbara (University Konstanz/GERMANY) 6(4):99
Moreno, Joanna (URS Corporation/USA) 6(2):239
Morgan, Scott (URS - Dames & Moore/USA) 6(7):321
Morrill, Pamela J. (Camp, Dresser, & McKee, Inc./USA) 6(2):113
Morris, Damon (ThermoRetec Corporation/USA) 6(7):1
Mortimer, Marylove (Mississippi State University/USA) 6(5):329
Mortimer, Wendy (Bell Canada/CANADA) 6(2):27; 6(6):185, 203, 211,
Mossing, Christian (Hedeselskabet/DENMARK) 6(2):11
Mossmann, Jean-Remi (Centre National de Recherche sur les Sites et Sols Pollués/FRANCE) 6(2):129
Moteleb, Moustafa A. (University of Cincinnati/USA) 6(6):133
Mowder, Carol S. (URS/USA) 6(7):321, 333, 341
Moyer, Ellen E. (ENSR International./USA) 6(1):75
Mravik, Susan C. (U.S. EPA/USA) 6(1):167
Mueller, James G. (URS Corporation/USA) 6(2):239
Müller, Axel (Water Technology Center/GERMANY) 6(8):105
Müller, Beate (Umweltschutz Nord GmbH/GERMANY) 6(4):131
Müller, Klaus (Battelle Europe/SWITZERLAND) 6(5):41; 6(6):241
Muniz, Herminio (Hart Crowser Inc./USA) 6(10):9
Murphy, Sean M. (Komex International Ltd./CANADA) 6(4):7
Murray, Cliff (United States Army Corps of Engineers/USA) 6(9):281
Murray, Gordon Bruce (Stella-Jones Inc./CANADA) 6(3):197

Murray, Willard A. (Harding ESE/USA) 6(7):197
Mutch, Robert D. (Brown and Caldwell/USA) 6(2):145
Mutti, Francois (Water & Soil Remediation S.r.l./ITALY) 6(6):179
Myasoedova, Nina M. (Russian Academy of Sciences/RUSSIA) 6(3):75

Nadolishny, Alex (Nedatek, Inc./USA) 6(10):139
Nagle, David P. (University of Oklahoma/USA) 6(5):61
Nam, Kyoungphile (Seoul National University/REPUBLIC OF KOREA) 6(9):195
Narayanaswamy, Karthik (Parsons Engineering Science/USA) 6(2):65
Nelson, Mark D. (Delta Environmental Consultants, Inc./USA) 6(1):175
Nelson, Yarrow (California Polytechnic State University/USA) 6(10):311
Nemati, M. (University of Cape Town/REP OF SOUTH AFRICA) 6(6):339
Nestler, Catherine C. (Applied Research Associates, Inc./USA) 6(4):59, 6(6):43
Nevárez-Moorillón, G.V. (UACH/MEXICO) 6(6):361
Neville, Scott L. (Aerojet General Corp./USA) 6(9):297
Newell, Charles J. (Groundwater Services, Inc./USA) 6(7):19
Nieman, Karl (Utah State University/USA) 6(4):67
Niemeyer, Thomas (Hochtief Umwelt Gmbh/GERMANY) 6(6):227
Nies, Loring (Purdue University/USA) 6(3):125
Nipshagen, Adri A.M. (IWACO/THE NETHERLANDS) 6(7):141
Nishino, Shirley (U.S. Air Force/USA) 6(3):59
Nivens, David E. (University of Tennessee/USA) 6(4):45
Noffsinger, David (Westinghouse Savannah River Company/USA) 6(10):163

Noguchi, Takuya (Nippon Institute of Technology/JAPAN) 6(5):321
Nojiri, Hideaki (The University of Tokyo/JAPAN) 6(4):51; 6(6):111
Noland, Scott (NESCO Inc./USA) 6(10):73
Nolen, C. Hunter (Camp Dresser & McKee/USA) 6(6):287
Norris, Robert D. (Eckenfelder/Brown and Caldwell/USA) 6(2):145; 6(7):35
North, Robert W. (Environ Corporation./USA) 6(7):189
Novak, John T. (Virginia Polytechnic Institute & State University/USA) 6(2):105; 6(5):1
Novick, Norman (Exxon/Mobil Oil Corp/USA) 6(1):35
Nuttall, H. Eric (The University of New Mexico/USA) 6(9): 155, 223
Nuyens, Dirk (Environmental Resources Management/BELGIUM) 6(7):87; 6(9):87
Nzengung, Valentine A. (University of Georgia/USA) 6(9):289

Ochs, L. Donald (Regenesis/USA) 6(10):139
O'Connell, Joseph E. (Environmental Resolutions, Inc./USA) 6(1):91
Odle, Bill (Newfields, Inc./USA) 6(5):189
O'Donnell, Ingrid (BAE Systems Properties, Ltd./UK) 6(6):119
Ogden, Richard (BAE Systems Properties Ltd./UK) 6(6):119
Oh, Byung-Taek (The University of Iowa/USA) 6(8):147, 175
Oh, Seok-Young (University of Delaware/USA) 6(6):149
Omori, Toshio (The University of Tokyo/JAPAN) 6(4):51; 6(6):111
O'Neal, Brenda (ARA/USA) 6(3):43
Oppenheimer, Carl H. (Oppenheimer Biotechnology/USA) 6(10):345
O'Regan, Gerald (Chevron Products Company/USA) 6(1):203
O'Reilly, Kirk T. (Chevron Research & Technology Co/USA) 6(1):83, 145, 203
Oshio, Takahiro (University of Tsukuba/JAPAN) 6(9):187

Ozdemiroglu, Ece (EFTEC Ltd./UK) 6(4):17

Padovani, Marco (Centro Ricerche Ambientali/ITALY) 6(4):131
Paganetto, A. (Istituto di Cibernetica e Biofisica/ITALY) 6(5):157
Pahr, Michelle R. (ARCADIS Geraghty & Miller/USA) 6(1):107
Pal, Nirupam (California Polytechnic State University/USA) 6(2):1
Palmer, Tracy (Applied Power Concepts, Inc./USA) 6(7):103
Palumbo, Anthony V. (Oak Ridge National Laboratory/USA) 6(4):73; 6(9):165
Panciera, Matthew A. (University of Connecticut/USA) 6(7):69
Pancras, Tessa (Wageningen University/THE NETHERLANDS) 6(5):289
Pardue, John H. (Louisiana State University/USA) 6(5): 207, 313; 6(9):273
Park, Kyoohong (ChoongAng University/REPUBLIC OF KOREA) 6(6):51
Parkin, Gene F. (University of Iowa/USA) 6(3):1
Paspaliaris, Ioannis (National Technical University of Athens/GREECE) 6(9):97
Paton, Graeme I. (Aberdeen University/UK) 6(1):1
Patrick, John (University of Reading/UK) 6(10):337
Payne, Frederick C. (ARCADIS Geraghty & Miller/USA) 6(7):53
Payne, Jo Ann (DuPont Co./USA) 6(8):43
Peabody, Jack G. (Regenesis/USA) 6(10):95
Peacock, Aaron D. (University of Tennessee/USA) 6(4):73; 6(5):305
Peargin, Tom R. (Chevron Research & Technology Co/USA) 6(1):67
Peeples, James A. (Metcalf & Eddy, Inc./USA) 6(7):173
Pehlivan, Mehmet (Tait Environmental Management, Inc./USA) 6(10):267, 275

Pelletier, Emilien (ISMER/CANADA) 6(2):57
Pennie, Kimberley A. (Stella-Jones, Inc./CANADA) 6(3):197
Peramaki, Matthew P. (Leggette, Brashears, & Graham, Inc./USA) 6(10):259
Perey, Jennie R. (University of Delaware/USA) 6(6):149
Perez-Vargas, Josefina (CINVESTAV-IPN/MEXICO) 6(6):219
Perina, Tomas (IT Corporation/USA) 6(1):51; 6(2):73
Perlis, Shira R. (Rowan University/USA) 6(5):215
Perlmutter, Michael W. (EnSafe, Inc./USA) 6(9):315
Perrier, Michel (École Polytechnique de Montréal/CANADA) 6(4):139
Perry, L.B. (U.S. Army Engineer Research & Development Center/USA) 6(5):9
Persico, John L. (Blasland, Bouck & Lee, Inc./USA) 6(10):319
Peschong, Bradley J. (Leggette, Brashears & Graham, Inc./USA) 6(1):175
Peters, Dave (URS/USA) 6(7):333
Peterson, Lance N. (North Wind Environmental, Inc./USA) 6(7):265
Petrovskis, Erik A. (Geotrans Inc./USA) 6(8):19
Peven-McCarthy, Carole (Battelle Ocean Sciences/USA) 6(5):231
Pfiffner, Susan M. (University of Tennessee/USA) 6(4):73
Phelps, Tommy J. (Oak Ridge National Laboratory/USA) 6(4):73
Pickett, Tim M. (Applied Biosciences Corporation/USA) 6(9):331
Pickle, D.W. (Equilon Enterprises LLC/USA) 6(8):61
Pierre, Stephane (École Polytechnique de Montréal/CANADA) 6(10):171
Pijls, Charles G.J.M. (TAUW BV/THE NETHERLANDS) 6(10):253
Pirkle, Robert J. (Microseeps, Inc./USA) 6(1):35
Pisarik, Michael F. (New Fields/USA) 6(1):121

Piveteau, Pascal (Institut Français du Pétrole/FRANCE) 6(1):153
Place, Matthew (Battelle/USA) 6(10):245
Plata, Nadia (Battelle Europe/SWITZERLAND) 6(5):41
Poggi-Varaldo, Hector M. (CINVESTAV-IPN/MEXICO) 6(3):243; 6(6):219
Pohlmann, Dirk C. (IT Corporation/USA) 6(2):253
Pokethitiyook, Prayad (Mahidol University/THAILAND) 6(10):329
Polk, Jonna (U.S. Army Corps of Engineers/USA) 6(9):281
Pope, Daniel F. (Dynamac Corp/USA) 6(1):129
Porta, Augusto (Battelle Europe/SWITZERLAND) 6(5):41; 6(6):241; 6(10):109
Portier, Ralph J. (Louisiana State University/USA) 6(5):305
Powers, Leigh (Georgia Institute of Technology/USA) 6(9):165
Prandi, Alberto (Water & Soil Remediation S.r.l/ITALY) 6(6):179
Prasad, M.N.V. (University of Hyderabad/INDIA) 6(5):165
Price, Steven (Camp Dresser & McKee, Inc./USA) 6(9):303
Priester, Lamar E. (Priester & Associates/USA) 6(10):65
Pritchard, P. H. (Hap) (U.S. Navy/USA) 6(7):125
Profit, Michael D. (CDM Federal Programs Corporation/USA) 6(6):81
Prosnansky, Michal (Gunma University/JAPAN) 6(9):201
Pruden, Amy (University of Cincinnati/USA) 6(1):19
Ptacek, Carol J. (University of Waterloo/CANADA) 6(9):71

Radosevich, Mark (University of Delaware/USA) 6(6):9
Radtke, Corey (INEEL/USA) 6(3):9
Raetz, Richard M. (Global Remediation Technologies, Inc./USA) 6(6):311
Rainwater, Ken (Texas Tech University/USA) 6(3):9

Ramani, Mukundan (University of Cincinnati/USA) 6(5):269
Raming, Julie B. (Georgia-Pacific Corp./USA) 6(1):183
Ramírez, N. E. (ECOPETROL-ICP/COLOMBIA) 6(6):319
Ramsay, Bruce A. (Polyferm Canada Inc./CANADA) 6(2):121; 6(10):171
Ramsay, Juliana A. (Queen's University/CANADA) 6(2):121; 6(10):171
Rao, Prasanna (University of Cincinnati/USA) 6(9):1
Ratzke, Hans-Peter (Umweltschutz Nord GMBH/GERMANY) 6(4):131
Reardon, Kenneth F. (Colorado State University/USA) 6(8):53
Rectanus, Heather V. (Virginia Polytechnic Institute & State University/USA) 6(2):105
Reed, Thomas A. (URS Corporation/USA) 6(8):157; 6(10):15, 95
Rees, Hubert (CEFAS Laboratory/UK) 6(10):337
Rehm, Bernd W. (RMT, Inc./USA) 6(2):97; 6(10):201
Reinecke, Stefan (Franz Environmental Inc./CANADA) 6(6):211
Reinhard, Martin (Stanford University/USA) 6(7):95
Reisinger, H. James (Integrated Science & Technology Inc/USA) 6(1):183
Rek, Dorota (IT Corporation/USA) 6(2):73
Reynolds, Charles M. (U.S. Army Engineer Research & Development Center/USA) 6(5):9
Reynolds, Daniel E. (Air Force Institute of Technology/USA) 6(2):173
Rice, John M. (RMT, Inc./USA) 6(7):181
Richard, Don E. (Barr Engineering Company/USA) 6(3):219; 6(5):105
Richardson, Ian (Conestoga-Rovers & Associates/USA) 6(10):131
Richnow, Hans H. (UFZ-Centre for Environmental Research/GERMANY) 6(4):99

Rijnaarts, Huub H.M. (TNO Institute of Environmental Science/THE NETHERLANDS) 6(2):231
Ringelberg, David B. (U.S. Army Corps of Engineers/USA) 6(5):9; 6(6):43; 6(10):115
Ríos-Leal, E. (CINVESTAV-IPN/MEXICO) 6(3):243
Ripp, Steven (University of Tennessee/USA) 6(4):45
Ritter, Michael (URS Corporation/USA) 6(2):239
Ritter, William F. (University of Delaware/USA) 6(6):9
Riva, Vanessa (Parsons Engineering Science, Inc./USA) 6(2):39
Rivas-Lucero, B.A. (Centro de Investigacion en Materiales Avanzados/MEXICO) 6(6):361
Rivetta, A. (Universita degli Studi di Milano/ITALY) 6(5):157
Robb, Joseph (ENSR International/USA) 6(1):75
Robertiello, Andrea (EniTecnologie S.p.A./ITALY) 6(6):173
Robertson, K. (Queen's University/CANADA) 6(2):121
Robinson, David (ERM, Inc./USA) 6(7):45
Robinson, Sandra L. (Virginia Polytechnic Institute & State University/USA) 6(5):1
Rockne, Karl J. (University of Illinois-Chicago/USA) 6(5):297
Rodríguez-Vázquez, Refugio (CINVESTAV-IPN/MEXICO) 6(3):243; 6(6):219, 369
Römkens, Paul (Alterra/THE NETHERLANDS) 6(5):137
Rongo, Rocco (University della Calabria/ITALY) 6(4):131
Roorda, Marcus L. (Rowan University/USA) 6(5):215
Rosser, Susan J. (University of Cambridge/UK) 6(5):69
Rowland, Martin A. (Lockheed-Martin Michoud Space Systems/USA) 6(7):1
Royer, Richard (The Pennsylvania State University/USA) 6(8):201

Ruggeri, Bernardo (Politecnico di Torino/ITALY) 6(3):211
Ruiz, Graciela M. (University of Iowa/USA) 6(1):195
Rupassara, S. Indumathie (University of Illinois at Urbana-Champaign/USA) 6(5):181
Sacchi, G.A. (Universita degli Studi di Milano/ITALY) 6(5):157
Sahagun, Tracy (U.S. Marine Corps./USA) 6(10):1
Sakakibara, Yutaka (Waseda University/JAPAN) 6(8):113; 6(9):201
Sakamoto, T. (Queen's University/CANADA) 6(10):171
Salam, Munazza (Crescent Heights High School/CANADA) 6(5):53
Salanitro, Joseph P. (Equilon Enterprises, LLC/USA) 6(1):11; 6(8):61
Salvador, Maria Cristina (CSD-GEOKLOCK/BRAZIL) 6(1):27
Samson, Réjean (École Polytechnique de Montréal/CANADA) 6(3):115; 6(4):139; 6(9):27
San Felipe, Zenaida (Monash University/AUSTRALIA) 6(4):1
Sánchez, F.N. (ECOPETROL-ICP/COLOMBIA) 6(6):319
Sánchez, Gisela (PDVSA Intevep/VENEZUELA) 6(6):257
Sánchez, Luis (PDVSA Intevep/VENEZUELA) 6(6):257
Sanchez, M. (Universidad Complutense de Madrid/SPAIN) 6(4):29; 6(6):377
Sandefur, Craig A. (Regenesis/USA) 6(7):257; 6(10):87
Sanford, Robert A. (University of Illinois at Urbana-Champaign/USA) 6(9):179
Santangelo-Dreiling, Theresa (Colorado Dept. of Transportation/USA) 6(10):231
Saran, Jennifer (Kennecott Utah Copper Corp./USA) 6(9):323
Sarpietro, M.G. (Universita di Catania/ITALY) 6(3):149

Sartoros, Catherine (Université du Québec à Montréal/CANADA) 6(3):165
Saucedo-Terán, R.A. (Centro de Investigacion en Materiales Avanzados/MEXICO) 6(6):361
Saunders, James A. (Auburn University/USA) 6(9):105
Sayler, Gary S. (University of Tennessee/USA) 6(4):45
Scalzi, Michael M. (Innovative Environmental Technologies, Inc./USA) 6(10):23
Scarborough, Shirley (IT Corporation/USA) 6(2):253
Schaffner, I. Richard (GZA GeoEnvironmental, Inc./USA) 6(7):165
Scharp, Richard A. (U.S. EPA/USA) 6(9):9
Schell, Heico (Water Technology Center/GERMANY) 6(8):105
Scherer, Michelle M. (The University of Iowa/USA) 6(3):1
Schipper, Mark (Groundwater Services) 6(8):73
Schmelling, Stephen (U.S. EPA/USA) 6(1):129
Schnoor, Jerald L. (University of Iowa/USA) 6(8):147
Schoefs, Olivier (École Polytechnique de Montréal/CANADA) 6(4):139
Schratzberger, Michaela (CEFAS Laboratory/UK) 6(10):337
Schulze, Susanne (Water Technology Center/GERMANY) 6(2):137
Schuur, Jessica H. (Lund University/SWEDEN) 6(6):65
Scrocchi, Susan (Conestoga-Rovers & Associates/USA) 6(1):99; 6(10):131
Sczechowski, Jeff (California Polytechnic State University/USA) 6(10):311
Seagren, Eric A. (University of Maryland/USA) 6(10):57
Sedran, Marie A. (University of Cincinnati/USA) 6(1):19
Seifert, Dorte (Technical University of Denmark/DENMARK) 6(2):11
Semer, Robin (Harza Engineering Company, Inc./USA) 6(7):157

Semprini, Lewis (Oregon State University/USA) 6(10):145, 155, 179
Seracuse, Joe (Harding ESE/USA) 6(7):205
Serra, Roberto (Centro Ricerche Ambientali/ITALY) 6(4):131
Sewell, Guy W. (U.S. EPA/USA) 6(1):167; 6(7):125; 6(8):167
Sharma, Pawan (Camp Dresser & McKee Inc./USA) 6(7):305
Sharp, Robert R. (Manhattan College/USA) 6(7):117
Shay, Devin T. (Groundwater & Environmental Services, Inc./USA) 6(10):101
Shelley, Michael L. (Air Force Institute of Technology/USA) 6(5):95
Shen, Hai (Dynamac Corporation/USA) 6(1): 129, 167
Sherman, Neil (Louisiana-Pacific Corporation/USA) 6(3):83
Sherwood Lollar, Barbara (University of Toronto/CANADA) 6(4):91, 109
Shi, Jing (EFX Systems, Inc./USA) 6(8):19
Shields, Adrian R.G. (Komex Europe/UK) 6(10):123
Shiffer, Shawn (University of Illinois/USA) 6(9):179
Shin, Won Sik (Lousiana State University/USA) 6(5):313
Shiohara, Kei (Mississippi State University/USA) 6(6):279
Shirazi, Fatemeh R. (Stratum Engineering Inc./USA) 6(8):121
Shoemaker, Christine (Cornell University/USA) 6(4):125
Sibbett, Bruce (IT Corporation/USA) 6(2):73
Silver, Cannon F. (Parsons Engineering Science, Inc./USA) 6(10):283
Silverman, Thomas S. (RMT, Inc./USA) 6(10):201
Simon, Michelle A. (U.S. EPA/USA) 6(10):293
Sims, Gerald K. (USDA-ARS/USA) 6(5):181
Sims, Ronald C. (Utah State University/USA) 6(4):67; 6(6):1
Sincock, M. Jennifer (ENVIRON International Corp./USA) 6(7):189

Sittler, Steven P. (Advanced Pollution Technologists, Ltd./USA) 6(2):215
Skladany, George J. (ERM, Inc./USA) 6(7):45, 213
Skubal, Karen L. (Case Western Reserve University/USA) 6(8):193
Slenders, Hans (TNO-MEP/THE NETHERLANDS) 6(7):289
Slomczynski, David J. (University of Cincinnati/USA) 6(2):247
Slusser, Thomas J. (Wright State University/USA) 6(5):95
Smallbeck, Donald R. (Harding Lawson/USA) 6(10):231
Smets, Barth F. (University of Connecticut/USA) 6(7):69
Smith, Christy (North Carolina State University/USA) 6(1):145
Smith, Colin C. (University of Sheffield/UK) 6(2):207
Smith, John R. (Alcoa Inc./USA) 6(6):89
Smith, Jonathan (The Environment Agency/UK) 6(4):17
Smith, Steve (King's College London/UK) 6(2):223; 6(3):173; 6(10):123
Smyth, David J.A. (University of Waterloo/CANADA) 6(9):71
Sobecky, Patricia (Georgia Institute of Technology/USA) 6(9):165
Sola, Adrianna (Spelman College/USA) 6(9):53
Sordini, E. (EniTechnologie/ITALY) 6(6):173
Sorensen, James A. (University of North Dakota/USA) 6(6):141
Sorenson, Kent S. (Idaho National Engineering and Environmental Laboratory./USA) 6(7):265
South, Daniel (Harding ESE/USA) 6(7):205
Spain, Jim (U.S. Air Force/USA) 6(3):59; 6(7):125
Spasova, Irena Ilieva (University of Mining & Geology/BULGARIA) 6(9):97
Spataro, William (University della Calabria/ITALY) 6(4):131

Spinnler, Gerard E. (Equilon Enterprises, LLC/USA) 6(1):11; 6(8):61
Springael, Dirk (VITO/BELGIUM) 6(4):35
Srinivasan, P. (GeoTrans, Inc./USA) 6(2):163
Stansbery, Anita (California Polytechnic State University/USA) 6(10):311
Starr, Mark G. (DuPont Co./USA) 6(8):43
Stehmeier, Lester G. (NOVA Research Technology Centre/CANADA) 6(4):117; 6(5):53
Stensel, H. David (University of Washington/USA) 6(10):211
Stordahl, Darrel M. (Camp Dresser & McKee Inc./USA) 6(6):287
Stout, Scott (Battelle/USA) 6(5):237
Strand, Stuart E. (University of Washington/USA) 6(10):211
Stratton, Glenn (Nova Scotia Agricultural College/CANADA) 6(3):197
Strybel, Dan (IT Corporation/USA) 6(9):215
Stuetz, R.M. (Cranfield University/UK) 6(6):329
Suarez, B. (ECOPETROL-ICP/COLOMBIA) 6(6):319
Suidan, Makram T. (University of Cincinnati/USA) 6(1):19; 6(5):243, 253, 261; 6(6):133,
Suthersan, Suthan S. (ARCADIS Geraghty & Miller/USA) 6(7):53
Suzuki, Masahiro (Nippon Institute of Technology/JAPAN) 6(5):321
Sveum, Per (Deconterra AS/NORWAY) 6(6):157
Swallow, Ian (BAE Systems Properties Ltd./UK) 6(6):119
Swann, Benjamin M. (Camp Dresser & McKee Inc./USA) 6(7):305
Swannell, Richard P.J. (AEA Technology Environment/UK) 6(10):337

Tabak, Henry H. (U.S. EPA/USA) 6(5):243, 253, 261, 269; 6(9):1, 17
Takai, Koji (Fuji Packing/JAPAN) 6(10):345

Talley, Jeffrey W. (University of Notre Dame/USA) 6(3):189; 6(4):59; 6(6):43; 6(7):125; 6(10):115
Tao, Shu (Peking University/CHINA) 6(7):61
Taylor, Christine D. (North Carolina State University/USA) 6(1):83
Ter Meer, Jeroen (TNO Institute of Environmental Science/THE NETHERLANDS) 6(2):231; 6(7):289
Tétreault, Michel (Royal Military College of Canada/CANADA) 6(8):95
Tharpe, D.L. (Equilon Enterprises LLC/USA) 6(8):61
Theeuwen, J. (Grontmij BV/THE NETHERLANDS) 6(7):289
Thomas, Hartmut (WASAG DECON GMbH/GERMANY) 6(3):67
Thomas, Mark (EG&G Technical Services, Inc./USA) 6(10):49
Thomas, Paul R. (Thomas Consultants, Inc./USA) 6(5):189
Thomas, Robert C. (University of Georgia/USA) 6(9):105
Thomson, Michelle M. (URS Corporation/USA) 6(2):81
Thornton, Steven F. (University of Sheffield/UK) 6(1):59, 6(2):207
Tian, C. (University of Cincinnati/USA) 6(8):35
Tiedje, James M. (Michigan State University/USA) 6(7):125; 6(8):19
Tiehm, Andreas (Water Technology Center/GERMANY) 6(2):137; 6(8):105
Tietje, David (Foster Wheeler Environmental Corportation/USA) 6(9):249
Timmins, Brian (Oregon State University/USA) 6(10):179
Togna, A. Paul (Envirogen Inc/USA) 6(9):281
Tolbert, David E.(U.S. Army/USA) 6(9):281
Tonnaer, Haimo (TAUW BV/THE NETHERLANDS) 6(7):297; 6(10):253
Toth, Brad (Harding ESE/USA) 6(10):231

Tovanabootr, Adisorn (Oregon State University/USA) 6(10):145
Travis, Bryan (Los Alamos National Laboratory/USA) 6(10):163
Trudnowski, John M. (MSE Technology Applications, Inc./USA) 6(9):35
Truax, Dennis D. (Mississippi State University/USA) 6(9):241
Trute, Mary M. (Camp Dresser & McKee, Inc./USA) 6(2):113
Tsuji, Hirokazu (Obayashi Corporation Ltd./JAPAN) 6(6):111, 249; 6(10):239
Tsutsumi, Hiroaki (Prefectural University of Kumamoto/JAPAN) 6(10):345
Turner, Tim (CDM Federal Programs Corp./USA) 6(6):81
Turner, Xandra (International Biochemicals Group/USA) 6(10):23
Tyner, Larry (IT Corporation/USA) 6(1):51; 6(2):73

Ugolini, Nick (U.S. Navy/USA) 6(10):65
Uhler, Richard (Battelle/USA) 6(5):237
Unz, Richard F. (The Pennsylvania State University/USA) 6(8):201
Utgikar, Vivek P. (U.S. EPA/USA) 6(9):17

Valderrama, Brenda (Universidad Nacional Autónoma de México/MEXICO) 6(6):17
Vallini, Giovanni (Universita degli Studi di Verona/ITALY) 6(3):267
van Bavel, Bert (Umeå University/SWEDEN) 6(3):181
van Breukelen, Boris M. (Vrije University/THE NETHERLANDS) 6(4):91
VanBroekhoven, K. (Catholic University of Leuven/BELGIUM) 6(4):35
Vandecasteele, Jean-Paul (Institut Français du Pétrole/FRANCE) 6(3):227
VanDelft, Frank (NOVA Chemicals/CANADA) 6(5):53
van der Gun, Johan (BodemBeheer bv/THE NETHERLANDS) 6(5):289

van der Werf, A. W. (Bioclear Environmental Technology/THE NETHERLANDS) 6(8):11
van Eekert, Miriam (TNO Environmental Sciences /THE NETHERLANDS) 6(2):231; 6(7):289
Van Hout, Amy H. (IT Corporation/USA) 6(3):35
Van Keulen, E. (DHV Environment and Infrastructure/THE NETHERLANDS) 6(8):11
Vargas, M.C. (ECOPETROL-ICP/COLOMBIA) 6(6):319
Vazquez-Duhalt, Rafael (Universidad Nacional Autónoma de México/MEXICO) 6(6):17
Venosa, Albert (U.S. EPA/USA) 6(1):19
Verhaagen, P. (Grontmij BV/THE NETHERLANDS) 6(7):289
Verheij, T. (DAF/THE NETHERLANDS) 6(7):289
Vidumsky, John E. (E.I. du Pont de Nemours & Company/USA) 6(2):81; 6(8):185
Villani, Marco (Centro Ricerche Ambientali/ITALY) 6(4):131
Vinnai, Louise (Investigative Science Inc./CANADA) 6(2):27
Visscher, Gerolf (Province of Groningen/THE NETHERLANDS) 6(7):141
Voegeli, Vincent (TranSystems Corporation/USA) 6(7):229
Vogt, Bob (Louisiana-Pacific Corporation/USA) 6(3):83
Volkering, Frank (TAUW bv/THE NETHERLANDS) 6(4):91
von Arb, Michelle (University of Iowa) 6(3):1
Vondracek, James E. (Ashland Inc./USA) 6(5):121
Vos, Johan (VITO/BELGIUM) 6(9):87
Voscott, Hoa T. (Camp Dresser & McKee, Inc./USA) 6(7):305
Vough, Lester R. (University of Maryland/USA) 6(5):77

Waisner, Scott A. (TA Environmental, Inc./USA) 6(4):59; 6(10):115

Walecka-Hutchison, Claudia M. (University of Arizona/USA) 6(9):231
Wall, Caroline (CEFAS Laboratory/UK) 6(10):337
Wallace, Steve (Lattice Property Holdings Plc./UK) 6(4):17
Wallis, F.M. (University of Natal/REP OF SOUTH AFRICA) 6(6):101; 6(9):79
Walton, Michelle R. (Idaho National Engineering & Environmental Laboratory/USA) 6(7):77
Walworth, James L. (University of Arizona/USA) 6(9):231
Wan, C.K. (Hong Kong Baptist University/CHINA) 6(6):73
Wang, Chuanyue (Rice University/USA) 6(5):85
Wang, Qingren (Chinese Academy of Sciences/CHINA [PRC]) 6(9):113
Wani, Altaf (Applied Research Associates, Inc./USA) 6(10):115
Wanty, Duane A. (The Gillette Company/USA) 6(7):87
Warburton, Joseph M. (Parsons Engineering Science/USA) 6(7):173
Watanabe, Masataka (National Institute for Environmental Studies/JAPAN) 6(5):321
Watson, James H.P. (University of Southampton/UK) 6(9):61
Wealthall, Gary P. (University of Sheffield/UK) 6(1):59
Weathers, Lenly J. (Tennessee Technological University/USA) 6(8):139
Weaver, Dallas E. (Scientific Hatcheries/USA) 6(1):91
Weaverling, Paul (Harding ESE/USA) 6(10):231
Weber, A. Scott (State University of New York at Buffalo/USA) 6(6):89
Weeber, Philip A. (Geotrans/USA) 6(10):163
Wendt-Potthoff, Katrin (UFZ Centre for Environmental Research/GERMANY) 6(9):43
Werner, Peter (Technical University of Dresden/GERMANY) 6(3):227; 6(8):105

West, Robert J. (The Dow Chemical Company/USA) 6(2):89
Westerberg, Karolina (Stockholm University/SWEDEN) 6(3):133
Weston, Alan F. (Conestoga-Rovers & Associates/USA) 6(1):99; 6(10):131
Westray, Mark (ThermoRetec Corp/USA) 6(7):1
Wheater, H.S. (Imperial College of Science and Technology/UK) 6(10):123
White, David C. (University of Tennessee/USA) 6(4):73; 6(5):305
White, Richard (EarthFax Engineering Inc/USA) 6(6):263
Whitmer, Jill M. (GeoSyntec Consultants/USA) 6(9):105
Wick, Lukas Y. (Swiss Federal Institute of Technology/SWITZERLAND) 6(3):251
Wickramanayake, Godage B. (Battelle/USA) 6(10):1
Widada, Jaka (The University of Tokyo/JAPAN) 6(4):51
Widdowson, Mark A. (Virginia Polytechnic Institute & State University/USA) 6(2):105; 6(5):1
Wieck, James M. (GZA GeoEnvironmental, Inc./USA) 6(7):165
Wiedemeier, Todd H. (Parsons Engineering Science, Inc./USA) 6(7):241
Wiessner, Arndt (UFZ - Centre for Environmental Research/GERMANY) 6(5):337
Wilken, Jon (Harding ESE/USA) 6(10):231
Williams, Lakesha (Southern University at New Orleans/USA) 6(5):145
Williamson, Travis (Battelle/USA) 6(10):245
Willis, Matthew B. (Cornell University/USA) 6(4):125
Willumsen, Pia Arentsen (National Environmental Research Institute/DENMARK) 6(3):141
Wilson, Barbara H. (Dynamac Corporation/USA) 6(1):129
Wilson, Gregory J. (University of Cincinnati/USA) 6(1):19

Wilson, John T. (U.S. EPA/USA) 6(1):43, 167
Wiseman, Lee (Camp Dresser & McKee Inc./USA) 6(7):133
Wisniewski, H.L. (Equilon Enterprises LLC/USA) 6(8):61
Witt, Michael E. (The Dow Chemical Company/USA) 6(2):89
Wong, Edwina K. (University of Guelph/CANADA) 6(6):185
Wong, J.W.C. (Hong Kong Baptist University/CHINA) 6(6):73
Wood, Thomas K. (University of Connecticut/USA) 6(5):199
Wrobel, John (U.S. Army/USA) 6(5):207

Xella, Claudio (Water & Soil Remediation S.r.l./ITALY) 6(6):179
Xing, Jian (Global Remediation Technologies, Inc./USA) 6(6):311

Yamamoto, Isao (Sumitomo Marine Research Institute/JAPAN) 6(10):345
Yamazaki, Fumio (Hyogo Prefectural Institute of Environmental Science/JAPAN) 6(5):321
Yang, Jeff (URS Corporation/USA) 6(2):239
Yerushalmi, Laleh (Biotechnology Research Institute/CANADA) 6(3):165
Yoon, Woong-Sang (Sam) (Battelle/USA) 6(7):13
Yoshida, Takako (The University of Tokyo/JAPAN) 6(4):51; 6(6):111
Yotsumoto, Mizuyo (Obayashi Corporation Ltd./JAPAN) 6(6):111
Young, Harold C. (Air Force Institute of Technology/USA) 6(2):173

Zagury, Gérald J. (École Polytechnique de Montréal/CANADA) 6(9): 27, 129
Zahiraleslamzadeh, Zahra (FMC Corporation/USA) 6(7):221
Zaluski, Marek H. (MSE Technology Applications/USA) 6(9):35
Zappi, Mark E. (Mississippi State University/USA) 6(9):241

Author Index

Zelennikova, Olga (University of Connecticut/USA) *6*(7):69
Zhang, Chuanlun L. (University of Missouri/USA) *6*(9):165
Zhang, Wei (Cornell University/USA) *6*(4):125
Zhang, Zhong (University of Nevada Las Vegas/USA) *6*(9):257
Zheng, Zuoping (University of Oslo/NORWAY) *6*(2):181
Zocca, Chiara (Universita degli Studi di Verona/ITALY) *6*(3):267
Zwick, Thomas C. (Battelle/USA) *6*(10):1

KEYWORD INDEX

This index contains keyword terms assigned to the articles in the ten-volume proceedings of the Sixth International In Situ and On-Site Bioremediation Symposium (San Diego, California, June 4-7, 2001). Ordering information is provided on the back cover of this book.

In assigning the terms that appear in this index, no attempt was made to reference all subjects addressed. Instead, terms were assigned to each article to reflect the primary topics covered by that article. Authors' suggestions were taken into consideration and expanded or revised as necessary. The citations reference the ten volumes as follows:

6(1): Magar, V.S., J.T. Gibbs, K.T. O'Reilly, M.R. Hyman, and A. Leeson (Eds.), *Bioremediation of MTBE, Alcohols, and Ethers*. Battelle Press, Columbus, OH, 2001. 249 pp.

6(2): Leeson, A., M.E. Kelley, H.S. Rifai, and V.S. Magar (Eds.), *Natural Attenuation of Environmental Contaminants*. Battelle Press, Columbus, OH, 2001. 307 pp.

6(3): Magar, V.S., G. Johnson, S.K. Ong, and A. Leeson (Eds.), *Bioremediation of Energetics, Phenolics, and Polycyclic Aromatic Hydrocarbons*. Battelle Press, Columbus, OH, 2001. 313 pp.

6(4): Magar, V.S., T.M. Vogel, C.M. Aelion, and A. Leeson (Eds.), *Innovative Methods in Support of Bioremediation*. Battelle Press, Columbus, OH, 2001. 197 pp.

6(5): Leeson, A., E.A. Foote, M.K. Banks, and V.S. Magar (Eds.), *Phytoremediation, Wetlands, and Sediments*. Battelle Press, Columbus, OH, 2001. 383 pp.

6(6): Magar, V.S., F.M. von Fahnestock, and A. Leeson (Eds.), *Ex Situ Biological Treatment Technologies*. Battelle Press, Columbus, OH, 2001. 423 pp.

6(7): Magar, V.S., D.E. Fennell, J.J. Morse, B.C. Alleman, and A. Leeson (Eds.), *Anaerobic Degradation of Chlorinated Solvents*. Battelle Press, Columbus, OH, 2001. 387 pp.

6(8): Leeson, A., B.C. Alleman, P.J. Alvarez, and V.S. Magar (Eds.), *Bioaugmentation, Biobarriers, and Biogeochemistry*. Battelle Press, Columbus, OH, 2001. 255 pp.

6(9): Leeson, A., B.M. Peyton, J.L. Means, and V.S. Magar (Eds.), *Bioremediation of Inorganic Compounds*. Battelle Press, Columbus, OH, 2001. 377 pp.

6(10): Leeson, A., P.C. Johnson, R.E. Hinchee, L. Semprini, and V.S. Magar (Eds.), *In Situ Aeration and Aerobic Remediation*. Battelle Press, Columbus, OH, 2001. 391 pp.

A

abiotic/biotic dechlorination **6(8)**:193
acenaphthene **6(5)**:253
acetate as electron donor **6(3)**:51; **6(9)**:297
acetone **6(2)**:49
acid mine drainage, (*see also* mine tailings) **6(9)**:1, 9, 27, 35, 43, 53
acrylic vessel **6(5)**:321
actinomycetes **6(10)**:211
activated carbon biomass carrier **6(6)**:311; **6(8)**:113

activated carbon **6(8)**:105
adsorption **6(3)**:243; **6(5)**:253; **6(6)**:377; **6(7)**:77; **6(8)**:131; **6(9)**:86
advanced oxidation **6(1)**:121; **6(10)**:33
aerated submerged **6(10)**:329
aeration **6(6)**:203
anaerobic/aerobic treatment **6(6)**:361; **6(7)**:229
age dating **6(5)**:231, 237
air sparging **6(1)**:115, 175; **6(2)**:239; **6(9)**:215; **6(10)**:1, 9, 41, 49, 65, 101, 115, 123, 163, 223
alachlor **6(6)**:9
algae **6(5)**:181
alkaline phosphatase **6(9)**:165
alkane degradation **6(5)**:313
alkylaromatic compounds **6(6)**:173
alkylbenzene **6(2)**:19
alkylphenolethoxylate **6(5)**:215
Amaranthaceae **6(5)**:165
Ames test **6(6)**:249
ammonia **6(1)**:175; **6(5)**:337
amphipod toxicity test **6(5)**:321
anaerobic **6(1)**:35, 43; **6(3)**:91; 205; **6(5)**:17, 25, 261, 297, 313; **6(6)**:133; **6(7)**:249, 297; **6(9)**:147, 303
anaerobic biodegradation **6(1)**:137; **6(5)**:1; **6(8)**:167
anaerobic bioventing **6(3)**:9
anaerobic petroleum degradation **6(5)**:25
anaerobic sparging **6(7)**:297
aniline **6(6)**:149
Antarctica **6(2)**:57
anthracene **6(3)**:165, 251; **6(6)**:73
aquatic plants **6(5)**:181
arid-region soils **6(9)**:231
aromatic dyes **6(6)**:369
arsenic **6(2)**:239, 261; **6(5)**:173; **6(9)**: 97, 129
atrazine **6(5)**:181; **6(6)**:9
azoaromatic compounds **6(6)**:149
Azomonas **6(6)**:219

B

bacterial transport **6(8)**:1
barrier technologies **6(1)**:11; **6(3)**:165; **6(7)**:289; **6(8)**:61, 79, 87, 105, 121; **6(9)**:27, 71, 195, 209, 309
basidiomycete **6(6)**:101
benthic **6(10)**:337

benzene **6(1)**:1, 67, 75, 145, 167, 203; **6(4)**:91,117; **6(8)**:87; **6(10)**:123
benzene, toluene, ethylbenzene, and xylenes (BTEX) **6(1)**:43, 51, 59, 107, 129, 167, 195; **6(2)**:11, 19, 137, 215, 223, 270; **6(4)**:99; **6(5)**:33; **6(7)**:133; **6(8)**:105; **6(10)**: 1, 23, 49, 65, 95, 123, 131
benzo(a)pyrene **6(3)**:149; **6(6)**:101
benzo(e)pyrene **6(3)**:149
BER, *see* biofilm-electrode reactor
bioassays **6(3)**:219
bioaugmentation **6(1)**:11; **6(3)**:133; **6(4)**:59; **6(6)**:9, 43, 111; **6(7)**:125; **6(8)**:1, 11, 19, 27, 43, 53, 61, 147, 175
bioavailability **6(3)**:115, 157, 173, 189, 51; **6(4)**:7; **6(5)**:253, 279, 289; **6(6)**:1
bioavailable FeIII assay **6(8)**:209
biobarrier **6(1)**:11; **6(3)**:165; **6(7)**:289; **6(8)**:61, 79, 105, 121; **6(9)**:27, 71, 209, 309
BIOCHLOR model **6(2)**:155
biocide **6(7)**:321, 333
biodegradability **6(6)**:193
biodegradation **6(1)**:19,153; **6(3)**:165, 181, 205, 235; **6(10)**:187
biofilm **6(3)**:251; **6(4)**:149; **6(8)**:79; **6(9)**:201, 303
biofilm-electrode reactor (BER) **6(9)**:201
biofiltration **6(4)**:149
biofouling **6(7)**:321, 333
bioindicators **6(1)**:1; **6(3)**:173; **6(5)**:223
biological carbon regeneration **6(8)**:105
bioluminescence **6(1)**:1; **6(3)**:173; **6(4)**:45
biopile **6(6)**:81, 127, 141, 227, 249, 287
bioreactors **6(1)**:91; **6(6)**:361; **6(8)**:11, 35; **6(9)**:1, 265, 281, 303, 315; **6(10)**:171, 211
biorecovery of metals **6(9)**:9
bioreporters **6(4)**:45
biosensors **6(1)**:1
bioslurping **6(10)**:245, 253, 267, 275
bioslurry and bioslurry reactors **6(3)**:189; **6(6)**:51, 65
biosparging **6(10)**:115, 163
biostabilization **6(6)**:89
biostimulation **6(6)**:43
biosurfactant **6(3)**:243; **6(7)**:53
bioventing **6(10)**:109, 115, 131
biphasic reactor **6(3)**:181

Keyword Index

biological oxygen demand (BOD) 6(10):311
BTEX, see benzene, toluene, ethylbenzene, and xylenes
Burkholderia cepacia 6(1):153; 6(7):117; 6(8):53
butane 6(1):137, 161
butyrate 6(7):289

C

cadmium 6(3):91; 6(9):79, 147
carAa, see carbazole 1,9a-dioxygenase gene
carbazole-degrading bacterium 6(6):111
carbazole 1,9a-dioxygenase gene (*carAa*) 6(4):51
Carbokalk 6(9):43
carbon isotope 6(4):91, 99, 109, 117; 6(10):115
carbon tetrachloride (CT) 6(2):81, 89; 6(5):113; 6(7):241; 6(8):185, 193
cesium-137 6(5):231
CF, see chloroform
charged coupled device camera 6(2):207
chelators addition (EDGA, EDTA) 6(5):129, 137, 145, 151; 6(9):123, 147
chemical oxidation 6(7):45
chicken manure 6(9):289
chlorinated ethenes 6(7):27, 61, 69, 109; 6(10):163, 201, 231
chlorinated solvents 6(2):145; 6(7):all; 6(8):19; 6(10):231
chlorobenzene 6(8):105
chloroethane 6(2):113; 6(7):133, 249
chloroform (CF) 6(2):81; 6(8):193
chloromethanes 6(8):185
chlorophenol 6(3):75, 133
chlorophyll fluorescence 6(5):223
chromated copper arsenate 6(9):129
chromium (Cr[VI]) 6(8):139, 147; 6(9):129, 139, 315
chrysene 6(6):101
citrate and citric acid 6(5):137; 6(7):289
cleanup levels 6(6):1
coextraction method 6(4):51
Coke Facility waste 6(2):129
combined chemical toxicity (*see also* toxicity) 6(5):305
cometabolic air sparging 6(10):145, 155, 223

cometabolism 6(1):137, 145, 153, 161; 6(2):19; 6(6):81, 141; 6(7):117; 6(10):145, 155, 163, 171, 179, 193, 201, 211, 217, 223, 231; 239
competitive inhibition 6(2):19
composting 6(3):83; 6(5):129, 6(6):73, 119, 165, 257; 6(7):141
constructed wetlands 6(5):173, 329
contaminant aging 6(3):157, 197
contaminant transport 6(3):115
copper 6(9):79, 129
cosolvent effects 6(1):175, 195, 203, 243
cosolvent extraction 6(7):125
cost analyses and economics of environmental restoration 6(1):129; 6(4):17; 6(8):121; 6(9):331; 6(10):65, 211
Cr(VI), see chromium
creosote 6(3):259; 6(4):59; 6(5):1, 237, 329; 6(6):81, 101, 141, 295
cresols 6(10):123
crude oil 6(5):313; 6(6):193, 249; 6(10):329
CT, see carbon tetrachloride
cyanide 6(9):331
cytochrome P-450 6(6):17

D

2,4-DAT, see diaminotoluene
DCA, see dichloroethane
1,1-DCA, see 1,1-dichloroethane
1,2-DCA, see 1,2-dichloroethane
DCE, see dichloroethene
1,1-DCE, see 1,1-dichloroethene
1,2-DCE, see 1,2-dichloroethene
c-DCE, see cis-dichloroethene
DCM, see dichloromethane
DDT, see also dioxins and pesticides 6(6):157
2,4-DNT, see dinitrotoluene
dechlorination kinetics 6(2):105; 6(7):61
dechlorination 6(2):231; 6(3):125; 6(5):95; 6(7):13, 61, 165, 173, 333; 6(8):19, 27, 43
DEE, see diethyl ether
Dehalococcoides ethenogenes 6(8):19, 43
dehalogenation 6(8):167
denaturing gradient gel electrophoresis (DGGE) 6(1):19; 6(4):35

denitrification **6(2)**:19; **6(4)**:149; **6(5)**:17, 261; **6(8)**:95; **6(9)**:179, 187, 195, 201, 209, 223, 309
dense, nonaqueous-phase liquid (DNAPL) **6(7)**:13, 19, 35, 181; **6(10)**:319
depletion rate **6(1)**:67
desorption **6(3)**:235, 243; **6(5)**:253; **6(6)**:377; **6(7)**:53, 77; **6(8)**:131
DGGE, *see* denaturing gradient gel electrophoresis
DHPA, *see* dihydroxyphenylacetate
dialysis sampler **6(5)**:207
diaminotoluene (2,4-DAT) **6(6)**:149
dibenzofuran-degrading bacterium **6(6)**:111
dibenzo-p-dioxin **6(6)**:111
dibenzothiophene **6(3)**:267
dichlorodiethyl ether **6(10)**:301
dichloroethane (DCA) **6(2)**:39; **6(7)**:289
1,1-dichloroethane (1,1-DCA; 1,2-DCA) **6(2)**:113; **6(5)**:207; **6(7)**:133, 165
1,2-dichloroethane (1,2-DCA) **6(5)**:207
dichloroethene, dichloroethylene **6(2)**:97, 155; **6(4)**:125; **6(5)**:105,113; **6(7)**:157, 197
cis-dichloroethene, *cis*-dichloroethylene (*c*-DCE) **6(2)**:39, 65, 73; 105, 173; **6(5)**:33, 95, 207; **6(7)**:1, 13, 61, 133, 141, 149, 165, 173, 181, 189, 205, 213, 221, 249, 273, 281, 289, 297, 305; **6(8)**:11, 19, 27, 43, 73, 105, 157, 209; **6(10)**:41, 145, 155, 179, 201
1,1-dichloroethene, 1,1-dichloroethylene (1,1-DCE) **6(2)**:39; **6(7)**:165, 229; **6(8)**:157; **6(10)**:231
1,2-dichloroethene and 1,2-dichloroethylene (1,2-DCE) **6(2)**:113
dichloromethane (DCM) **6(2)**:81; **6(8)**:185
diesel fuel **6(1)**:175; **6(2)**:57; **6(5)**:305; **6(6)**:81, 141, 165; **6(10)**:9
diesel-range organics (DRO) **6(10)**:9
diethyl ether (DEE) **6(1)**:19
dihydroxyphenylacetate (DHPA) **6(4)**:29
diisopropyl ether (DIPE) **6(1)**:19, 161
1,3-dinitro-5-nitroso-1,3,5-triazacyclohexane (MNX) (*see also* explosives *and* energetics) **6(3)**:51; **6(8)**:175
dinitrotoluene (2,4-DNT) **6(3)**:25, 59; **6(6)**:127, 149
dioxins **6(6)**:111

DIPE, *see* diisopropyl ether
dissolved oxygen **6(2)**:189, 207
16S rDNA sequencing **6(8)**:19
DNAPL, *see* dense, nonaqueous-phase liquid
DNX, *see* explosives and energetics
DRO, *see* diesel-range organics
dual porosity aquifer **6(1)**:59
dyes **6(6)**:369

E

ecological risk assessment **6(4)**:1
ecotoxicity, (*see also* toxicity) **6(1)**:1; **6(4)**:7
ethylenedibromide (EDB) **6(10)**:65
EDGA, *see* chelate addition
EDTA, *see* chelate addition
effluent **6(4)**:1
electrokinetics **6(9)**:241, 273
electron acceptors and electron acceptor processes **6(2)**:1, 137, 163, 231; **6(5)**:17, 25, 297; **6(7)**:19
electron donor amendment **6(3)**:25, 35, 51, 125; **6(7)**:69, 103,109, 141, 181, 249, 289, 297; **6(8)**:73; **6(9)**:297, 315
electron donor delivery **6(7)**:19, 27, 133, 173, 213, 221, 265, 273, 281, 305
electron donor mass balance **6(2)**:163
electron donor transport **6(4)**:125; **6(7)**:133; **6(9)**:241
embedded carrier **6(9)**:187
encapsulated bacteria **6(5)**:269
enhanced aeration **6(10)**:57
enhanced desorption **6(7)**:197
environmental stressors **6(4)**:1
enzyme induction **6(6)**:9; **6(10)**:211
ERIC sequences **6(4)**:29
ethane **6(2)**:113; **6(7)**:149
ethanol **6(1)**:19,167,175, 195, 203; **6(5)**:243; **6(6)**:133; **6(9)**:289
ethene and ethylene **6(2)**:105,113; **6(5)**:95; **6(7)**:1, 95, 133, 141, 205, 281, 297, 305; **6(8)**:11, 43, 167, 175, 209
ethylene dibromide **6(10)**:193
explosives and energetics **6(3)**:9, 17, 25, 35, 43, 51, 67; **6(5)**:69; **6(6)**:119, 127, 133; **6(7)**:125

F

fatty acids **6(5)**:41
Fe(II), *see* iron
Fenton's reagent **6(6)**:157
fertilizer **6(5)**:321; **6(6)**:35; **6(10)**:337
fixed-bed and fixed-film reactors **6(5)**:221, 337; **6(6)**:361; **6(9)**:303
flocculants **6(6)**:279
flow sensor **6(10)**:293
fluidized-bed reactor **6(1)**:91; **6(6)**:133, 311; **6(9)**:281
fluoranthene **6(3)**:141; **6(6)**:101
fluorogenic probes **6(4)**:51
food safety **6(9)**:113
formaldehyde **6(6)**:329
fractured shale **6(10)**:49
free-product recovery **6(6)**:211
Freon **6(2)**:49
fuel oil **6(5)**:321
fungal remediation **6(3)**:75, 99; **6(5)**:61, 279; **6(6)**:17, 101, 157, 263, 319, 329, 369
Funnel-and-Gate™ **6(8)**:95

G

gas flux **6(6)**:185
gasoline **6(1)**:35, 75, 161, 167, 195; **6(10)**:115
gasoline-range organics (GRO) **6(10)**:9
manufactured gas plants and gasworks **6(2)**:137; **6(10)**:123
GCW, *see* groundwater circulating well
gel-encapsulated biomass **6(8)**:35
GEM, *see* genetically engineered microorganisms
genetically engineered microorganisms (GEM) **6(4)**:45; **6(5)**:199; **6(7)**:125
genotoxicity, (*see also* toxicity) **6(3)**:227
Geobacter **6(3)**:1
geochemical characterization **6(4)**:91
geographic information system (GIS) **6(2)**:163
geologic heterogeneity **6(2)**:11
germination index 6(3):219; **6(6)**:73
GFP, *see* green fluorescent protein
GIS, *see* geographic information system
glutaric dialdehyde dehydrogenase **6(4)**:81
Gordonia terrae **6(1)**:153
green fluorescent protein (GFP) **6(5)**:199
GRO, see gasoline-range organics

groundwater **6(3)**:35; **6(8)**: 35, 87, 121; **6(10)**:231
groundwater circulating well (GCW) **6(7)**:229, 321; **6(10)**:283, 293

H

H_2 gas, *see* hydrogen
H_2S, *see* hydrogen sulfide
halogenated hydrocarbons **6(9)**:61
halorespiration **6(8)**:19
heavy metal **6(2)**:239; **6(5)**:137, 145, 157, 165, 173; **6(6)**:51; **6(9)**:53, 61, 71, 79, 86, 97, 113, 129, 147
herbicides **6(5)**:223; **6(6)**:35
hexachlorobenzene **6(3)**:99
hexane **6(3)**:181, **6(6)**:329
HMX, *see* explosives and energetics
hollow fiber membranes **6(5)**:269
hopane **6(6)**:193; **6(10)**:337
hornwort **6(5)**:181
HRC® (a proprietary hydrogen-release compound) **6(3)**:17, 25, 107; **6(7)**:27, 103, 157, 189, 197, 205, 221, 257, 305, **6(8)**:157, 209
^{2}H-tetradecane (*see also* tetradecane) **6(2)**:27
humates **6(1)**:99
hybrid treatment **6(10)**:311
hydraulic containment **6(8)**:79
hydraulically facilitated remediation **6(2)**:239
hydrocarbon **6(6)**:235; **6(10)**:329
hydrogen (H_2 gas) **6(2)**:199; **6(9)**:201
hydrogen injection, in situ **6(7)**:19
hydrogen isotope **6(4)**:91
hydrogen peroxide **6(1)**:121; **6(6)**:353; **6(10)**:33
hydrogen release compound, *see* HRC®
hydrogen sulfide (H_2S) **6(9)**:123
hydrogen **6(2)**:231, **6(7)**:61, 305
hydrolysis **6(1)**:83
hydrophobicity **6(3)**:141
hydroxyl radical **6(1)**:121
hydroxylamino TNT intermediates **6(5)**:85

I

immobilization **6(8)**:53
immobilized cells **6(8)**:121
immobilized soil bioreactor **6(10)**:171

in situ oxidation 6(7):1
industrial effluents 6(6):303, 361
inhibition 6(9):17
injection strategies, in situ 6(7):19, 133, 173, 213, 221, 265, 273, 305, 313; 6(9):223; 6(10):23, 163
insecticides 6(6):27
intrinsic biodegradation 6(2):89, 121
intrinsic remediation, see natural attenuation
ion migration 6(9):241
iron (Fe[II]) 6(5):1
iron barrier 6(8):139, 147, 157, 167
iron oxide 6(3):1
iron precipitation 6(3):211
iron-reducing processes 6(2):121; 6(3):1; 6(5):1, 17, 25; 6(6):149; 6(8):193, 201, 209; 6(9):43, 323
IR-spectroscopy 6(4):67
isotope analyses 6(2):27; 6(4):91; 6(8):27
isotope fractionation 6(4):99, 109, 117

J

jet fuel 6(10):95, 139

K

KB-1 strain 6(8):27
kerosene 6(6):219
kinetics 6(8):131, 6(1):1, 19, 27, 167; 6(2):11, 19, 105; 6(3):173; 6(4):131; 6(7):61
Klebsiella oxytoca 6(7):117
Kuwait 6(6):249

L

laccase 6(3):75; 6(6):319
lactate and lactic acid 6(7):103, 109, 165, 181, 213, 265, 281, 289; 6(8):139; 6(9):155, 273
lagoons 6(6):303
land treatment units (LTU) 6(6):1; 6(6):81, 141, 287, 295
landfarming 6(3):259; 6(4):59; 6(5):53, 279; 6(6):1, 43, 59, 179, 203, 211, 235
landfills 6(2):145, 247; 6(4):91; 6(8):113
leaching 6(9):187
lead 6(5):129, 145, 151, 157

lead-210 6(5):231
light, nonaqueous-phase liquids (LNAPL) 6(1):59; 6(4):35; 6(10):57, 109, 245, 253, 275
lindane, (*see also* pesticides) 6(5):189
linuron (*see also* herbicides) 6(5):223
LNAPL, *see* light, nonaqueous-phase liquids
Lolium multiflorum 6(5):9
LTU, *see* land treatment units
lubricating oil 6(6):173
luciferase 6(3):133
lux 6(4):45

M

mackinawite 6(9):155
macrofauna 6(10):337
magnetic separation 6(9):61
magnetite 6(3):1; 6(8):193
manganese 6(2):261
manufactured gas plant (MGP) 6(2):19; 6(3):211, 227; 6(10):123
mass balance 6(2):163
mass transfer limitation 6(3):157
mass transfer 6(1):67
MC-100, see mixed culture
media development 6(9):147
Meiofauna 6(5):305; 6(10):337
membrane 6(5):269; 6(9):1, 265
metabolites 6(3):227
metal reduction 6(8):1
metal precipitation 6(9):9, 165
metals, biorecovery of 6(9):9
metals speciation 6(9):129
metal toxicity (*see also* toxicity) 6(9):17, 129
metals 6(5):129, 305; 6(8):1; 6(9):9, 17, 27, 105, 123, 129, 155, 165
methane oxidation 6(10):171, 187, 193, 201, 223, 231
methane 6(1):183; 6(8):113
methanogenesis 6(1):35, 43, 183; 6(3):205; 6(9):147
methanogens 6(3):91
methanol 6(1):183; 6(7):141, 289, 297
methanotrophs 6(10):171, 187, 201
methylene chloride 6(2):39; 6(10):231
Methylosinus trichosporium 6(10):187
methyl *tert*-butyl ether *or* methyl *tertiary*-butyl ether (MTBE) 6(1):1, 11, 19, 27, 35, 43, 51, 59, 67, 75, 83, 91, 107,

115, 121, 129, 137, 145, 153,161, 195, **6(2)**:215; **6(8)**:61; **6(10)**:1, 65
MGP, *see* manufactured gas plant
microbial heterogeneity **6(4)**:73
microbial isolation **6(3)**:267
microbial population dynamics **6(4)**:35
microbial regrowth **6(2)**:253; **6(7)**:1, 13; **6(10)**:319
microcosm studies **6(7)**:109; **6(10)**:179
microencapsulation **6(8)**:53
microfiltration **6(9)**:201
microporous membrane **6(9)**:265
microtox assay **6(3)**:227
mine tailings (*see also* acid mine drainage) **6(5)**:173; **6(9)**:27, 71
mineral oil **6(5)**:279, 289; **6(6)**:59
mineralization **6(2)**:121; **6(3)**:165; **6(6)**:165; **6(8)**:175; **6(9)**:139, 155
MIP, *see* membrane interface probe
mixed culture **6(8)**:61
mixed wastes **6(3)**:91; **6(7)**:133; **6(9)**:139
MNX, *see* 1,3-dinitro-5-nitroso-1,3,5-triazacyclohexane
modeling **6(1)**:51; **6(2)**:105, 155, 181, **6(4)**:125, 131, 139, 149; **6(6)**:339, 377; **6(8)**:185; **6(9)**:27, 105; **6(10)**:163
moisture content **6(2)**:247
molasses as electron donor **6(3)**:35; **6(7)**:53, 103, 149, 173; **6(9)**:315
monitored natural attenuation (*see also* natural attenuation) **6(1)**:183, **6(2)**:11, 163, 199, 223, 253, 261
monitoring techniques **6(2)**:27,189, 199, 207; **6(4)**:59
motor oil **6(5)**:53
MPE, *see* multiphase extraction
multiphase extraction (MPE) well design **6(10)**:245, 259
MTBE, *see* methyl *tert*-butyl ether
multiphase extraction **6(10)**:245, 253, 259, 267, 275
municipal solid waste **6(2)**:247
Mycobacterium sp. IFP 2012 **6(1)**:153
Mycobacterium adhesion **6(3)**:251
mycoremediation **6(6)**:263

N

naphthalene **6(1)**:1; **6(2)**:121; **6(3)**:173, 227; **6(5)**:1, 253; **6(6)**:51; **6(8)**:95, **6(9)**:139; **6(10)**:123

NAPL, *see* nonaqueous-phase liquid
natural attenuation **6(1)**:27, 35, 43, 51, 59, 75, 83, 183, 195; **6(2)**:1,39, 73, 81, 89, 97, 105, 137, 145, 173, 181, 215; **6(4)**:91, 99, 117; **6(5)**:33, 189, 321; **6(8)**:185, 209; **6(9)**:179; **6(10)**:115, 163
natural gas **6(10)**:193
natural organic carbon **6(2)**:261
natural organic matter **6(2)**:81, 97; **6(8)**:201
natural recovery **6(5)**:132, 231
nitrate contamination **6(9)**:173
nitrate reduction **6(3)**:51; **6(5)**:25; **6(9)**:331
nitrate utilization efficiency **6(6)**:353
nitrate **6(2)**:1; **6(3)**:17, 43; **6(6)**:353; **6(8)**:95, 147; **6(9)**:179, 187, 195, 209, 223, 257
nitrification **6(4)**:149; **6(5)**:337; **6(9)**:215
nitroaromatic compounds (*see also* explosives and energetics) **6(3)**:59, 67; **6(6)**:149
nitrobenzene, *see also* explosives and energetics **6(6)**:149
nitrocellulose, *see also* explosives and energetics **6(6)**:119
nitrogen fixation **6(6)**:219
nitrogen utilization **6(9)**:231
nitrogenase **6(6)**:219
nitroglycerin, *see also* explosives and energetics **6(5)**:69
nitrotoluenes, *see also* explosives and energetics **6(6)**:127
nitrous oxide **6(8)**:113
^{13}C-NMR, *see* nuclear magnetic resonance spectroscopy
nonaqueous-phase liquids (NAPLs) **6(1)**:67, 203; **6(3)**:141; **6(7)**:249
nonylphenolethoxylates **6(5)**:215
nuclear magnetic resonance spectroscopy (^{13}C-NMR) **6(4)**:67
nutrient augmentation **6(3)**:59; **6(5)**:329; **6(6)**:257; **6(7)**:313; **6(9)**:331; **6(10)**:23
nutrient injection **6(10)**:101
nutrient transport **6(9)**:241

O

oily waste **6(4)**:35; 6(6):257; **6(10)**:337, 345
oil-coated stones **6(10)**:329

optimization **6(5)**:279
ORC® (a proprietary oxygen-release compound) **6(1)**:99,107; **6(2)**:215; **6(3)**:107; **6(7)**:229; **6(10)**:9, 15, 87, 95, 139
organic acids **6(2)**:39
organophosphorus **6(6)**:17, 27
advanced oxidation **6(6)**:157, **6(10)**:311
oxygen-release compound, *see* ORC®
oxygen-release material **6(10)**:73
oxygen respiration **6(9)**:231; **6(10)**:57
oxygenation **6(1)**:107, 145
ozonation **6(1)**:121; **6(10)**:33, 149, 301

P

packed-bed reactors **6(9)**:249; **6(10)**:329
PAHs, *see* polycyclic aromatic hydrocarbons
paper mill waste **6(4)**:1
paraffins **6(3)**:141
partitioning **6(9)**:129
PCBs, *see* polychlorinated biphenyls
PCP toxicity (*see also* toxicity) **6(3)**:125
PCP, *see* pentachlorophenol
PCR analysis, *see* polymerase chain reaction
pentachlorophenol (PCP) **6(3)**:83, 91, 99, 107, 115, 125; **6(5)**:329; **6(6)**:279, 287, 295, 329
percarbonate **6(10)**:73
perchlorate **6(9)**:249, 257, 265, 273, 281, 289, 297, 303, 309, 315
perchloroethene, perchloroethylene **6(7)**:53
permeable reactive barriers **6(3)**:1; **6(8)**: 73, 87, 95, 121, 139, 147, 157, 167, 175, 185; **6(9)**:71, 309, 323; **6(10)**:95
pesticides **6(5)**:189; **6(6)**:9, 17, 35
PETN reductase **6(5)**:69
petroleum hydrocarbon degradation **6(4)**:7; **6(5)**:9, 17, 25; **6(8)**:131; **6(10)**: 65, 101, 245, 345
phenanthrene **6(2)**:121; **6(3)**:227, 235, 243; **6(6)**:51, 65, 73
phenol **6(6)**:303, 319, 329
phenolic waste **6(6)**:311
phenol-oxidizing cultures **6(10)**:211, 217, 239
phenyldodecane **6(2)**:27
phosphate precipitation **6(9)**:165
PHOSter **6(10)**:65

photocatalysis **6(10)**:311
physical/chemical pretreatment **6(1)**:1, 51; **6(2)**:253; **6(3)**:149; **6(5)**:9, 33, 41, 53, 61, 69, 77, 85,105, 113, 121, 129,137, 145, 151, 157, 165, 189, 199, 207, 279, 337; **6(6)**:59, 157, 241; **6(7)**:1, 13; **6(9)**:113, 173; **6(10)**:239, 311, 319
phytotoxicity (*see also* toxicity) **6(5)**:41, 223
phytotransformation **6(5)**:85
pile-turner **6(6)**:249
PLFA, *see* phospholipid fatty acid analysis
polychlorinated biphenyls (PCBs) **6(2)**:39,105,173; **6(5)**:33, 61, 95, 113, 231, 289; **6(6)**:89, **6(7)**:13, 61, 69, 95, 109, 125, 133, 141, 149, 165, 181, 189, 197, 205, 213, 241, 249, 273, 297, 305; **6(8)**:11,19, 27, 43, 157, 167, 193, 209; **6(10)**:33, 41, 231, 283
polycyclic aromatic hydrocarbons (PAHs) **6(2)**:19, 121, 129, 137; **6(3)**:141, 149, 157, 165, 173, 181, 189, 197, 205, 211, 219, 227, 235, 243; **6(4)**:35, 45, 59, 67; **6(5)**:1, 9, 17, 41, 237, 243, 251, 253, 261, 269, 279, 289, 305, 329; **6(6)**:43, 51, 59, 65, 73, 81, 89, 101, 279, 295, 297; **6(7)**:125; **6(8)**:95; **6(9)**:139; **6(10)**:33, 123
polymerase chain reaction (PCR) analysis **6(4)**:29, 35, 51; **6(8)**:43
polynuclear aromatic hydrocarbons, *see* polycyclic aromatic hydrocarbons
poplar lipid fatty acid analysis (PLFA) **6(3)**:189
poplar trees **6(5)**:113, 121, 189
potassium permanganate **6(2)**:253; **6(7)**:1
precipitation **6(9)**:105; **6(10)**:301
pressurized-bed reactor **6(6)**:311
propane utilization **6(1)**:137; **6(10)**:145, 155, 179, 193
propionate **6(7)**:265, 289
Pseudomonas fluorescens **6(3)**:173
pyrene **6(3)**:165, 235; **6(4)**:67; **6(6)**: 65, 73, 101
pyridine **6(4)**:81

R

RABITT, *see* reductive anaerobic biological in situ treatment technology
radium **6(5)**:173
rapeseed oil **6(6)**:65
RDX, *see* research development explosive
rebound **6(10)**:1
recirculation well **6(7)**:333, 341; **6(10)**:283
redox measurement and control **6(1)**:35; **6(2)**:11, 231; **6(5)**:1; **6(9)**:53
reductive anaerobic biological in situ treatment technology (RABITT) **6(7)**:109
reductive dechlorination **6(2)**:39, 65, 97, 105, 145, 173; **6(4)**:125; **6(7)**:45, 53, 87, 103, 109, 133, 141, 149, 157,181, 197, 205, 213, 221, 249, 257, 265, 273, 289, 297; **6(8)**:11, 73, 105, 157, 209
reductive dehalogenation **6(7)**:69
reed canary grass **6(5)**:181
research development explosive (RDX) **6(3)**:1, 9, 17, 25, 35, 43, 51; **6(6)**:133; **6(8)**:175
respiration and respiration rates **6(2)**:129; **6(4)**:59; **6(6)**:185, 227
respirometry **6(6)**:127; **6(10)**:217
rhizoremediation **6(5)**:9, 61, 199
Rhodococcus opacus **6(4)**:81
risk assessment **6(2)**:215; **6(4)**:1
16S rRNA sequencing **6(8)**:43; **6(9)**:147
rock-bed biofiltration **6(4)**:149
rotating biological contactor **6(9)**:79
rototiller **6(6)**:203
RT3D **6(10)**:163

S

salinity **6(9)**:257
salt marsh **6(5)**:313
SC-100, *see* single culture
Sea of Japan **6(5)**:321
sediments **6(3)**:91; **6(5)**:231, 237, 253, 261, 269, 279, 289, 297, 305; **6(6)**:51, 59; **6(9)**:61
selenium **6(9)**:323, 331
semivolatile organic carbon (SVOC) **6(2)**:113
sheep dip **6(6)**:27
Shewanella putrefaciens **6(8)**:201
silicon oil **6(3)**:141, 181
single culture **6(8)**:61
site characterization **6(10)**:139
site closure **6(2)**:215
slow-release fertilizer **6(2)**:57
sodium glycine **6(9)**:273
soil treatment **6(3)**:181; **6(6)**:1
soil washing **6(5)**:243; **6(6)**:241
soil-vapor extraction (SVE) **6(1)**:183; **6(10)**:1, 41, 131, 223
solids residence time **6(10)**:211
sorption **6(5)**:215, 253; **6(6)**:377; **6(8)**:131; **6(9)**:79, 105
source zone **6(7)**:13, 19, 27, 181; **6(10)**:267
soybean oil **6(7)**:213
sparging **6(10)**:33, 145, 155
stabilization **6(6)**:89
substrate delivery **6(7)**:281
sulfate reduction **6(1)**:35; **6(3)**:43, 91; **6(5)**:261, 313; **6(6)**:339; **6(7)**:69, 95; **6(8)**:139, 147, 193; **6(9)**:1, 9, 17, 27, 35, 43, 61, 71, 86, 105, 123, 147
sulfide precipitation **6(9)**:123
surfactants **6(5)**:215; **6(6)**:73; **6(7)**:213, 321, 333; **6(8)**:131
sustainability **6(6)**:1
SVE, *see* soil vapor extraction
SVOC, *see* semivolatile organic carbon
synthetic pyrethroid **6(6)**:27

T

TCA, see trichlorethane
1,1,1-TCA, *see* 1,1,1-trichloroethane
1,1,2-TCA, *see* 1,1,2-trichloroethane
2,4,6-TCP, *see* 2,4,6-trichlorophenol
1,1,1,2-TeCA,*see* tetrachloroethane
1,1,2,2-TeCA, *see* tetrachloroethane
1,3,5-TNB, *see* 1,3,5-trinitrobenzene
TAME, *see* tertiary methyl-amyl ether
TBA, *see* tertiary butyl alcohol
TBF, *see* tertiary butyl formate
TCE oxidation, *see* trichloroethene, trichloroethylene
TCE, *see* trichloroethene
TCP, *see* trichlorophenol
t-DCE, *see* trans-dichloroethene, trans-dichloroethylene
technology comparisons **6(7)**:45; **6(9)**:323
terrazyme **6(10)**:345

tertiary butyl alcohol (TBA) **6(1)**:19, 27, 35, 51, 59, 91, 145, 153, 161
tertiary butyl formate (TBF) **6(1)**:145, 161
tertiary methyl-amyl ether (TAME) **6(1)**:59, 161
tetrachloroethane (1,1,1,2-TeCA, 1,1,2,2-TeCA) **6(5)**:207; **6(7)**:321, 341; **6(8)**:193
tetradecane (see also ^{2}H-tetradecane) **6(3)**:181
thermal desorption **6(3)**:189, **6(6)**:35
TNB, *see* trinitrobenzene
TNT, see trinitrotoluene
TNX, *see* 1,3,5-trinitroso-1,3,5-triazacyclohexane
tobacco plant **6(5)**:69
toluene **6(1)**:145; **6(2)**:181; **6(7)**:95; **6(8)**:35, 131
total petroleum hydrocarbons (TPH) **6(2)**:1; **6(5)**:9; **6(6)**:127, 173, 179, 193, 227, 241, 249; **6(10)**:15, 73, 115, 337
toxicity **6(1)**:1; **6(3)**:67, 189, 227; **6(4)**:7; **6(5)**:41, 61, 223, 305; **6(9)**:17, 129
TPH, *see* total petroleum hydrocarbons
trace gas emissions **6(6)**:185
trans-dichloroethene, trans-dichloroethylene **6(5)**:95, 207; **6(7)**:165
transgenic plants **6(5)**:69
transpiration **6(5)**:189
Trecate oil spill **6(6)**:241; **6(10)**:109
trichloroethane (TCA) **6(7)**:241, 281
1,1,1-trichloroethane (1,1,1-TCA; 1,1,2-TCA) **6(2)**:39, 113, 464; **6(5)**:207; **6(7)**:87,165, 281
1,1,2-trichloroethane (1,1,2-TCA) **6(5)**:207
trichloroethene, trichloroethylene (TCE) **6(2)**:39, 65, 73, 97, 105, 113, 155, 173, 253; **6(4)**:125; **6(5)**:33, 95, 105, 113, 207; **6(7)**:1, 13, 53, 61, 69, 77, 87, 109, 117, 133, 141, 149, 157, 181, 189, 197, 205, 213, 221, 241, 249, 265, 273, 281, 297, 305, **6(8)**:11, 19, 27, 35, 43, 53, 73, 105,147, 157, 193, 209; **6(10)**:41, 131, 145, 155, 163, 171, 179, 187, 201, 211, 217, 223, 231, 239, 283, 319
2,4,6-trichlorophenol (2,4,6-TCP) **6(3)**:75; **6(8)**:121
trichlorotrifluoroethane **6(2)**:49

trinitrobenzene (TNB) **6(3)**:9, 25
1,3,5-trinitroso-1,3,5-triazacyclohexane (TNX) **6(8)**:175
trinitrotoluene (TNT) **6(3)**:35, 67; **6(5)**:69, 77, 85; **6(6)**:133

U

underground storage tank (UST) **6(1)**:67, 129
uranium **6(5)**:173; **6(7)**:77; **6(9)**:155, 165
UST, *see* underground storage tank

V

vacuum extraction **6(1)**:115
vadose zone **6(1)**:183; **6(2)**:39, 65, 97, 105, 113, 155, 173; **6(3)**:9; **6(5)**:33, 105; **6(7)**:1,13, 61, 133, 141, 197, 205, 213, 249, 273, 281, 305; **6(8)**:11,19, 43, 73, 157, 209; **6(10)**:41, 163
vegetable oil **6(6)**:65; **6(7)**103, 213, 241, 249
vinyl chloride **6(2)**:73; **6(4)**:109; **6(5)**:95; **6(7)**:95,149, 157, 165, 173, 289, 297, **6(10)**:231
vitamin B_{12} **6(7)**:321, 333, 341
VOCs, *see* volatile organic carbons
volatile fatty acid **6(7)**:61
volatile organic carbons (VOCs) **6(2)**:113, 189; **6(5)**:113, 121

W

wastewater treatment **6(5)**:215; **6(6)**:149; **6(9)**:173
water potential **6(9)**:231
weathering **6(4)**:7
wetlands **6(5)**:33, 95, 105, 313, 329; **6(9)**:97
white rot fungi, (*see also* fungal remediation) **6(3)**:75, 99; **6(6)**:17, 157, 263
windrow **6(6)**:81, 119, 141
wood preservatives **6(3)**:83, 259; **6(4)**:59; **6(6)**:279

X

xylene **6(1)**:67

Y
yeast extract *6(7)*:181

Z
zero-valent iron *6(8)*:157, 167; *6(9)*:71
zinc *6(4)*:91; *6(9)*:79